Patenting Biotechnical Innovation

To Richard, Victoria and Thomas.

Patenting Biotechnical Innovation

Eligibility, Ethics and Public Interest

Edited by

Naomi Hawkins

Professor of Intellectual Property Law, School of Law, University of Sheffield, UK

Edward Elgar
PUBLISHING

Cheltenham, UK • Northampton, MA, USA

Cover image: JJ Ying on Unsplash.

Published by
Edward Elgar Publishing Limited
The Lypiatts
15 Lansdown Road
Cheltenham
Glos GL50 2JA
UK

Edward Elgar Publishing, Inc.
William Pratt House
9 Dewey Court
Northampton
Massachusetts 01060
USA

A catalogue record for this book
is available from the British Library

Library of Congress Control Number: 2022937607

This book is available electronically in the **Elgar**online
Law subject collection
http://dx.doi.org/10.4337/9781800884410

Printed on elemental chlorine free (ECF)
recycled paper containing 30% Post-Consumer Waste

ISBN 978 1 80088 440 3 (cased)
ISBN 978 1 80088 441 0 (eBook)

Printed and bound in the USA

Contents

Contributors

Mateo Aboy, Centre for Law, Medicine, and Life Sciences (LML), Faculty of Law, University of Cambridge, UK

Naomi Hawkins, School of Law, University of Sheffield, UK

Phoebe Li, School of Law, Politics and Sociology, University of Sussex, UK

Kathleen Liddell, Centre for Law, Medicine, and Life Sciences (LML), Faculty of Law, University of Cambridge, UK

Johnathon Liddicoat, Centre for Law, Medicine, and Life Sciences (LML), Faculty of Law, University of Cambridge, UK

Pheh Hoon Lim, School of Law, Auckland University of Technology, New Zealand

Aisling McMahon, Department of Law, Maynooth University, Republic of Ireland

Dianne Nicol, Centre for Law and Genetics University of Tasmania, Australia

Jane Nielsen, Centre for Law and Genetics, University of Tasmania, Australia

Ella O'Sullivan, Department of Law & Criminal Justice, Waterford Institute of Technology, Republic of Ireland

Andrew Sagar, Centre for Law, Medicine, and Life Sciences (LML), Faculty of Law, University of Cambridge, UK

Alison Slade, Leicester Law School, University of Leicester, UK

Karen Walsh, Law School, University of Exeter, UK

Acknowledgements

I am grateful to Professor Susan Kelly, Professor Charlotte Waelde, Professor Jane Kaye, Professor Dianne Nicol, Professor Liz Trinder, Professor Rebecca Probert, Professor Andrea Lista, Dr Mathilde Pavis, Dr Karen Walsh, Dr Richard Wassall and members of the Science, Culture and the Law research group and colleagues at the University of Exeter Law School for their valuable comments and support throughout this project.

The project of which this book was part was supported by the UK Economic and Social Research Council (grant code ES/K009575/1).

1. Introduction – patenting biotechnical innovation: in the public interest?

Naomi Hawkins

1. PATENTS AND BIOTECHNICAL INNOVATION: THE ROLE OF PATENT LAW

As the first body of law to address new inventions, patent law has often had to adapt to changes in technology. However, with the advent of recent innovations such as biotechnology, the nature of innovation has changed so much that principles of patent law have arguably been strained. Developments in technology are pushing the boundaries of what is possible, and are poised to have a revolutionary impact on science, technology, and ultimately everyday life. Biotechnology, genomics, gene editing, artificial intelligence and space technologies, among other areas, are all developing apace. Advances in biology mean that entirely synthetic organisms can be created in laboratories. Modifications to the germ line of organisms are possible. Laboratory fabrication of human body parts is now beginning to be achievable. Artificial intelligence is such that the role of human intelligence in the inventive process may shortly become less essential.

As these innovative fields grow, so too does the potential for improving the lives of people the world over. Innovative medicines can serve to cure or treat previously intractable disease burden. Prenatal testing can identify diseases before birth, allowing earlier intervention. Sophisticated and less invasive diagnostic tests can identify cancer at an earlier stage, enabling earlier, less aggressive and more effective treatment. Genetic diseases can be treated through gene therapy while antibodies promise the selective treatment of benign and disease-causing conditions. However, the expense of such treatments or interventions is often high, with the costs of care beyond the reach of patients and public health systems in high-income, middle-income and low-income countries alike.[1] Access for marginalised populations is even

[1] UCL Institute for Innovation and Public Purpose, 'The People's Prescription: Re-Imaging Health Innovation to Deliver Public Value' (2018); M Mazzucato, 'High

poorer – those without public access to healthcare, those without health insurance and especially those in the global south are all excluded by a system of innovation which results in extremely high prices and limited availability for innovative treatments.[2] Patents are often the reason that such monopoly prices are assured, and while they are not the sole reason for international access to medicine concerns, they are a significant contributor.

Understanding the interaction between technological advances and patent law and policy is important for two key reasons. First, patent law and policy have a significant impact on rates of innovation, which is key to addressing important and intractable challenges for our world.[3] Second, patent law and policy impact public access to the fruits of these innovations.[4] Patents exist because their incentive function theoretically serves the public interest,[5] but at the same time, patent policy can operate contrary to other important public interests such as access to medicine.

This balance between private rights and the public interest is fundamental to the patent system. A patent right is only justified if the benefits of that monopoly to the public outweigh the constraints posed by it. Although the patent

---

Cost of New Drugs' (2016) 354 British Medical Journal i4136; S Vincent Rajkumar, 'The High Cost of Prescription Drugs: Causes and Solutions' (2020) 10 Blood Cancer Journal 71.

2 See, for example: Lasantha Ariyarathna and Kanchana Kariyawasam, 'Pharmaceutical Patents and Access to Generic Medicines in Developing Countries' (2020) 42(2) EIPR 108; Siva Thambisetty, 'Improving Access to Patented Medicines: Are Human Rights Getting in the Way?' (2019) 4 IPQ 284; Olasupo Owoeye and Oluwabusayo Owoeye, 'Intellectual Property, Access to Medicines and Universal Health Coverage through a Health Right Lens' (2018) 40(1) EIPR 49; Thaddeus Manu, 'Essential Medicines and the Complexity of Implementing Nationally Based Compulsory Licensing: On the Need for a Regional System of Compulsory Licensing in Sub-Saharan Africa' (2014) 36 EIPR 39; Suerie Moon, Jorge Bermudez and Ellen 't Hoen, 'Innovation and Access to Medicines for Neglected Populations: Could a Treaty Address a Broken Pharmaceutical R&D System?' (2012) 9(5) PLoS Medicine e1001218; Ellen 't Hoen, Jonathan Berger, Alexandra Calamy and Suerie Moon, 'Driving a Decade of Change: HIV/AIDS, Patents and Access to Medicines for All' (2011) 14 Journal of the International AIDS Society 15.

3 Justine Pila, *The Subject Matter of Intellectual Property* (Oxford University Press 2017). For a discussion of the justifications of the intellectual property system see for example Robert P Merges, *Justifying Intellectual Property* (Harvard University Press 2011).

4 Rebecca Tushnet, 'Intellectual Property as a Public Interest Mechanism' in Rochelle Cooper Dreyfuss and Justine Pila (eds), *The Oxford Handbook of Intellectual Property Law* (Oxford University Press 2018).

5 Geertrui van Overwalle, 'Human Rights' Limitations in Patent Law' in FW Grosheide (ed), *Intellectual Property and Human Rights: A Paradox* (Edward Elgar Publishing 2010).

bargain is a fundamental unpinning and a long entrenched aspect of the patent system, the question of whether this balance is struck at the right level, and whether the public interest is adequately safeguarded, remains. This question is at the core of this book.

In order to enable this patent bargain, patents should be very much a background issue for the development and delivery of a new technology, especially in the health field – they should facilitate but not obstruct. When patents become central to the public discussion of a new technology, when patents become the focus of the story, then the patent system is not functioning as it should. It is then reasonable to turn our attention to the patent bargain and to the underlying justifications for the existence of patents, and where the lines are drawn around patentability and enforcement, to ensure that the right balance is achieved between those key features of the patent system – innovation and the public interest.

It is thus necessary that patent law and policy be studied to ensure that it operates to advance innovation in the public interest. This is particularly pressing in some key areas, such as biotechnology, where innovations have such important impact and ethical implications. The Covid-19 crisis has again brought into sharp relief the important role that the patent system plays in access to medicine. This crisis has demonstrated the need for collaborative solutions to access dilemmas, but also the ways in which collective action and social and political pressure can drive change.

This book examines patent law and policy in light of the patent bargain and the public interest. It considers the ways in which the public interest is recognised and safeguarded, as well as the ways in which the law fails to measure up in this regard. It adds to the scholarship in this area by drawing out the key aspects of the public interest across the full life cycle of the patent, from patent grant and validity, through considerations of the patentable subject matter and eligibility as well as exceptions and exclusions as well as patent licensing. The book does so in light of the scientific developments in new fields of biotechnology, and in light of recent case law.

1.1 The Patent Bargain and the Public Interest

Patents, and intellectual property (IP) rights more generally, are economic tools which operate within the market. Most economists, and those who subscribe to the law and economics school, maintain the presumption that profit-seeking market outcomes tend to be socially optimal.[6] It is generally accepted that

---

[6] Iain M Cockburn, 'O Brave New Industry, That Has Such Patents In It!' in F Scott Kieff (ed), *Perspectives on Properties of the Human Genome Project*, (Elsevier/Academic Press 2003).

the incentive function of patents is a primary justification for their existence, and that this incentive function must be balanced against the limitations on access which patents impose. The economic literature focuses particularly on the question of incentives.[7] However, the extent to which patents provide an incentive is highly contested, and the lack of effective incentive provided by patents for areas of unmet health needs is well established.[8] There is greater incentive to develop new treatments for chronic diseases of the wealthy than to undertake work in the areas of disease prevention or vaccines. Similarly, the development of much needed drugs to address problems of antimicrobial resistance is not adequately incentivised by the current system of drug development, which is structured around the development of products for ongoing treatment rather than short-term or single dose regimes.[9] The disputed nature of the incentive function of patents is well documented in other literature,[10] and is not the primary focus of this book. Instead, this book focuses on the patent bargain – the ways in which the balance is struck between patent monopoly rights and the public interest. The foundational nature of the patent bargain, and the manner in which it underpins patent law in the UK, has been reaffirmed in recent case law.[11] This book sets out to highlight the second part of the patent bargain: the focus of much of the literature in the field is on incentives, but those incentives are balanced against the restrictions that they pose *in the public interest*, and these questions of public interest deserve greater attention.

---

[7] Dominique Guellec and Bruno van Pottelsberghe de la Potterie, *The Economics of the European Patent System: IP Policy for Innovation and Competition* (Oxford University Press 2007).

[8] See for example T Reynolds, 'Investing in Prevention: What Incentive Does the Pharmaceutical Industry Have?' (2002) 94 Journal of the National Cancer Institute 1736; UCL Institute for Innovation and Public Purpose (n 1); Henry G Grabowski, Joseph A DiMasi and Genia Long, 'The Roles of Patents and Research and Development Incentives in Biopharmaceutical Innovation' (2015) 34 Health Affairs 302; Joseph E Stiglitz, 'Scrooge and Intellectual Property Rights' (2006) 333 British Medical Journal 1279.

[9] Matthew H Todd and others, 'There Is No Market for New Antibiotics: This Allows an Open Approach to Research and Development' (2021) 6 Wellcome Open Research 146.

[10] See for example Suerie Moon, Jorge Bermudez and Ellen 't Hoen (n 2); World Health Assembly (WHA) (2008) Global Strategy and Plan of Action on Public Health, Innovation and Intellectual Property. Resolution 61.21 (Sixty-first). www.who.int/gb/ebwha/pdf_files/A61/A61_R21-en.pdf; Robin C Feldman and others, 'Negative Innovation: When Patents Are Bad for Patients' (2021) 39 Nature Biotechnology 914.

[11] *Regeneron Pharmaceuticals Inc v Kymab* [2020] UKSC 27, [23]; *Warner-Lambert v Generics (UK) Ltd t/a Mylan* [2018] UKSC 56, [17].

The traditional approach to the characterisation of the public interest in patent law is similarly predominantly economic in focus.[12] Patents are a market tool, and problems arising from their operation, including concerns around access to patented technologies, are often viewed in terms of market failure. However, a focus on market failure tends to focus on economic questions, which limits the consideration of the public interest to a similarly narrow economic calculus. This has been particularly evident in the debates around questions of modern biomedical innovation, especially as regards pricing and access.[13] The limitations of this approach in the context of the Covid-19 crisis have been clear – reducing the public interest in these debates to a balancing of monetary and economic incentives, with scant attention paid to interests which are less easily reduced into monetary terms, is problematic. Increasingly, concerns are raised about the potential for IP monopolies to extract rents rather than to incentivise productive innovation for the benefit of society.[14]

In order to better frame the public interest in the balance of patent law and policy, it is important to look beyond this narrow economic calculus and expand its conception to its other components. Beyond only economics, the public interest encompasses concepts of *ordre public*, morality, consumer protection, safety, public health and regulation.[15] At present, there is no coherent or well-developed concept of the public interest which is universally applied in patent law. Current case law is focused primarily on the public benefit occasioned by the full disclosure of the invention, which brings with it the ability to work the invention after the expiry of the monopoly,[16] although more recent cases are beginning to highlight the wider aspects mentioned here.[17]

In addition, there is no single point in the life cycle of a patent at which the public interest is addressed. Instead, questions of the public interest can and do arise at multiple points in the patent life cycle, from the point of application, to grant, to exploitation and licensing and infringement. Concerns associated

---

12 van Overwalle (n 5).

13 Rochelle Cooper Dreyfuss, 'Patents and Human Rights: Where Is the Paradox?' in FW Grosheide (ed), *Intellectual Property and Human Rights: A Paradox* (Edward Elgar Publishing 2010); Stiglitz (n 8).

14 See for example Mariana Mazzucato, *The Value of Everything: Making and Taking in the Global Economy* (Penguin 2018).

15 van Overwalle (n 5).

16 *Regeneron Pharmaceuticals Inc v Kymab* [2020] UKSC 27, [23]; *Actavis v ICOS* [2019] UKSC 15 [53]; *Warner-Lambert v Generics (UK) Ltd t/a Mylan* [2018] UKSC 56 [17]; *EXXON/Fuel Oils (T 409/91)* [1994] OJ EPO 653, [3.3] and [3.4].

17 See for example *Evalve Inc v Edwards Lifesciences Ltd* [2020] EWHC 513 (Pat); *Edwards Lifesciences LLC v Boston Scientific Scimed Inc* [2018] EWHC 1256 (Pat), where public health was recognised as relevant, although the scope for its consideration is very narrow.

with the public interest are evident in relation to questions of patent eligibility and scope, questions of morality and ethics, and the way in which patents are exploited, including prices, and voluntary and compulsory licensing. The response to these public interest concerns is typically dealt with in a different manner by the courts and by legal scholars depending on the point at which the concern arises, and there is little or no connection made between these areas of scholarship. However, while there is little cohesiveness or logical connection between approaches to the public interest in patent law, there are common threads which can be discerned. This book seeks to make connections between different areas and approaches to questions of the public interest in patent law, through the lens of patents for emerging biotechnologies.

Accordingly, this book considers the role of the public interest across the patent lifespan. It begins by exploring questions of patent eligibility and the ways in which this affects both the scope of what is patentable, influencing the incentives offered by patent law, and correspondingly protecting the public interest, both through providing incentives and through limited effects on access. It goes on to explore the ways in which ethical and social questions, including the morality provisions as well as other legal controls on licensing, can support access to innovation, in the public interest. The following section explores these areas in more detail.

1.2 Patent Eligibility

Important and intractable questions continue to arise around questions of patent eligibility. Some of these difficulties relate to how the patentability criteria as traditionally conceived can fit with new types of technologies. Such questions are often considered at grant stage, and may be further scrutinised in subsequent litigation concerning the validity of the patent in question.

What constitutes an invention (in European parlance) or patentable subject matter (in the USA) has been the focus of important recent case law, in terms of both product and method patents.[18] From the patentability of genetic material to methods of diagnosis, case law to delineate the boundaries of what properly forms the subject matter which can be protected by a patent has explicitly focused on the protection of important public interests around access

---

[18] See for example *Mayo Collaborative Services v Prometheus Laboratories* Inc 566 US 66 (2012); *Association for Molecular Pathology v Myriad Genetics Inc* 569 US 576 (2013); *Alice Corp v CLS Bank International* 573 US 208 (2014); *D'Arcy v Myriad Genetics Inc* [2015] HCA 35; Dianne Nicol, et al, 'International Divergence in Gene Patenting' (2019) 20 Annual Review of Genomics and Human Genetics 519; Rochelle Dreyfuss, Jane Nielsen and Dianne Nicol, 'Patenting Nature – A Comparative Perspective' (2018) 6 Journal of Law and the Biosciences 1.

to medicine. Patent eligibility for cutting-edge technologies and products in biomedicine is considered in chapters of this book focused on precision medicine, non-invasive prenatal testing (NIPT) and biofabrication.[19]

The patentability criteria tend to receive less focus in discussions of the public interest. However, the boundaries of what falls within or outside of patentable inventions, being sufficiently novel, inventive, industrially applicable and sufficiently described, also have implications for the patent bargain. Recent case law challenging NIPT patents was argued in the UK around these more traditional requirements, in contrast to the USA and Australia, where the focus was instead on the questions of the scope of patentable subject matter.[20] But patentability criteria are also underpinned by legislative determinations of the balance of the public interest, and their application can widen or narrow the scope of what is or is not patentable in way which has important implications for the public interest. This connection of the requirements of patentability to the underlying patent bargain has been re-emphasised in recent UK Supreme Court cases,[21] with the potential for the reasoning of the court to be adopted in relation to the patentability criteria to embed the public interest more firmly in patent jurisprudence.[22]

1.3 Exceptions

Exceptions also provide limitations to the patentability of certain categories of inventions, with important public interest implications. European patent law, in contrast to some other jurisdictions, makes explicit provision for morality, whereby patents are not granted for inventions the subject matter of which is contrary to *ordre public* or morality.[23] In addition, there are specific provisions

---

[19] Andrew Sagar, Kathleen Liddell, Johnathon Liddicoat and Mateo Aboy, 'Patent Eligibility and Precision Medicine: Emerging Technologies, Legal Flux and Empirical Methods' in Naomi Hawkins (ed), *Patenting Biotechnical Innovation: Eligibility, Ethics and Public Interest* (Edward Elgar Publishing 2022); Dianne Nicol and Jane Nielsen, 'Non-invasive Prenatal Testing and the Resilience of the Patent System' in Naomi Hawkins (ed), *Patenting Biotechnical Innovation: Eligibility, Ethics and Public Interest* (Edward Elgar Publishing 2022); Pheh Hoon Lim and Phoebe Li, 'Patentability of Biofabricated Human Organs: "Products of Nature" or "Products Derived from Nature" Revisited' in Naomi Hawkins (ed), *Patenting Biotechnical Innovation: Eligibility, Ethics and Public Interest* (Edward Elgar Publishing 2022).

[20] Dianne Nicol and Jane Nielsen (n 19).

[21] *Generics (UK) Ltd (trading as Mylan) v Warner-Lambert Co LLC* [2018] UKSC 56 [17]; *Actavis v ICOS* [2019] UKSC 15 [54].

[22] Justine Pila, 'Reflections on a Post-Pandemic European Patent System' (2020) 42 European Intellectual Property Review 530.

[23] Art 53(a) European Patent Convention.

against the patentability of processes for cloning human beings, processes for modifying the germ line genetic identity of human beings, uses of human embryos for industrial or commercial purposes and processes for modifying the genetic identity of animals which are likely to cause them suffering without any substantial medical benefit.[24] There are clear public interest justifications for these provisions.[25] However, the extent to which the interpretations of these provisions serve to protect the public interest, or whether they are limited to too narrow a sphere to be effective in practice, continues to be debated.[26]

Questions of the public interest are often implicit in the analysis here. It is generally accepted that the legislature has already weighed the relevant considerations and determined where the line should be drawn in terms of the public interest.[27] In the vast majority of determinations of these tests, questions of the public interest will be very much in the background. However, both the ways in which fine distinctions are drawn and how tests are interpreted can have important implications for the public interest in cases at the margins. Wide interpretation of patent scope and narrow construction of exceptions results in differences in terms of both the incentives for patent holders and the costs of access. Moreover, the extent to which analysis is affected by considerations of the appropriate balance between incentives and the public interest is not always transparent.

1.4 Patent Exploitation

Once a patent is granted, there is still a great deal of scope for the public interest to feature in its exploitation. Much of the legal consideration around the public interest is focused around the validity of a patent, and whether a patent is granted is obviously important and significant. However, refusal of patent grant is not the only means by which the public interest can, or should, be protected. A granted patent may be exploited in a manner which enhances public

---

[24] Rule 28(a)–(d) of the EPC 2000 Implementing Regulations.

[25] Karen Walsh and Naomi Hawkins, 'Expanding the Role of Morality and Public Policy in European Patent Law' in Paul Torremans (ed) *Intellectual Property and Human Rights* (Wolters Kluwer 2020).

[26] These questions are explored in this book in Ella O'Sullivan 'A problem of interpretation: Is Article 53(a) EPC necessarily restrictive of emerging technologies?' in Naomi Hawkins (ed), *Patenting Biotechnical Innovation: Eligibility, Ethics and Public Interest* (Edward Elgar Publishing 2022).

[27] Karen Walsh, 'Invoking the Public Interest in the European Patent System' in Naomi Hawkins (ed), *Patenting Biotechnical Innovation: Eligibility, Ethics and Public Interest* (Edward Elgar Publishing 2022). See also discussion of the role of the public interest in patent law in *Chiron Corporation v Organon Teknika (No 10)* [1995] RPC 325.

welfare, or in a manner in which the public interest is overridden. The life of a patent following its grant (or a failure of any challenge to its validity) has huge potential to impact the public interest, and it is therefore important that considerations of the ethical and social aspects of patent law take into account factors relevant to exploitation, infringement, defences and licensing.

Concerns in this respect often focus on access to medicine, including the problem of excessive pricing of patented drugs, particularly in the developing world.[28] However, while pricing is an important question, questions of access to patented technologies are multifaceted and intractable, and pricing is only one aspect. Solutions to maximise the public interest can be found in three main areas: enforcement and defences to infringement; mechanisms associated with voluntary licensing; and non-voluntary licensing.

First, existing law offers some limited defences to infringement which can serve to enhance the public interest. The most notable of these in the UK context is probably the research exemption found in s60(5)(b) Patents Act 1977 UK. This research exception provides a defence to patent infringement for acts done for experimental purposes relating to the subject matter of the invention. The research exception is an established defence in most European jurisdictions,[29] but it is not clearly defined or uniformly applied across jurisdictions. Although the research exemption provides a vital means by which research and development on or into patented technologies can continue in the public interest, it will not extend to enabling the use of patented technologies outside of that research, and therefore it will not, for example provide a means for treatment with a new drug outside of the research context.[30]

Another defence which has limited practical effect at present – but which has the potential to be developed further – is the extemporaneous preparation of medicaments exception, found in s60(5)(c) of the UK Patents Act and in corresponding provisions in many European jurisdictions.[31] This provision sets out a defence to infringement for the extemporaneous preparation in a pharmacy of a medicine for an individual in accordance with a prescription

---

[28] See for example Cynthia Ho, *Access to Medicine in the Global Economy: International Agreements on Patents and Related Rights* (Oxford University Press 2011); Ruth Okediji and Margo Bagley, *Patent Law in Global Perspective* (Oxford University Press 2014).

[29] Art 27(b) Community Patent Convention.

[30] T Cook, A European Perspective as to the Extent to which Experimental Use, and Certain Other Defences to Patent Infringement, Apply to Differing Types of Research: A Report for the Intellectual Property Institute (Intellectual Property Institute, London 2006) 133–4; F Bor, 'Exemptions to Patent Infringement Applied to Biotechnology Research Tools' (2006) 28 EIPR 510.

[31] Similar provisions exist in a number of European jurisdictions, and the corresponding provision of the Community Patent Convention is Art 27(c).

given by a registered medical or dental practitioner. There is a public interest justification for this defence; it serves to protect pharmacists in carrying out their professional role and to ensure that patients obtain medical treatment from medical practitioners, including pharmacists, without obstruction caused by the necessity of having regard to patents.[32] An analogous defence for the preparation of genetic diagnostic tests has been proposed elsewhere.[33]

Public interest considerations can also play a role in questions of enforcement, although under current interpretations of the relevant provisions this role is extremely limited in practice. Recent case law demonstrates a potential, if limited, role, for the stay of injunctions on public health grounds in the public interest.[34] There is also at least a theoretical possibility of the application of Crown use as a defence, although again, the scope is very limited.[35] An injunction is a discretionary remedy, but the exercise of the discretion to refuse an injunction for patent infringement is unlikely without strong reasons, with the public interest being seen to be properly protected by the requirements of the Patents Act 1977.[36] Such discretion will be limited in practice to very unusual circumstances, and not of wide application. It is certainly unlikely that it could be used to reduce the costs of appropriate and available, but expensive, patented inventions.

Second, voluntary licensing is founded on the principle of freedom of contract – an inherently private right, which at first glance offers scant support for the invocation of the public interest. However, a number of important limitations on the untrammelled exercise of private rights exist in the patent context, which can serve to support the public interest. Existing mechanisms such as patent pools are designed to leverage combined bargaining power to reduce excessive pricing.[37] Recent efforts to encourage voluntary pledges of

---

32 B Domeij, *Pharmaceutical Patents in Europe* (Stockholm Studies in Law, Kluwer Law International, The Hague 2000) 310.

33 Naomi Hawkins, 'An Exception to Infringement for Genetic Testing – Addressing Patient Access and Divergence Between Law and Practice' (2012) 43 International Review of Intellectual Property and Competition Law 641.

34 *Evalve Inc v Edwards Lifesciences Ltd* [2020] EWHC 513 (Pat); *Edwards Lifesciences LLC v Boston Scientific Scimed Inc* [2018] EWHC 1256 (Pat).

35 *Ipcom GmbH & Co KG v Vodafone Group Plc* [2021] EWCA Civ 205, overturning *IPCom GmbH & Co KG v Vodafone Group Plc* [2020] EWHC 132 (Pat).

36 *Chiron Corporation v Organon Teknika (No 10)* [1995] RPC 325.

37 Ellen 't Hoen, *Private Patents and Public Health: Changing Intellectual Property Rules for Access to Medicines* (2016); S Soini, S Ayme and G Matthijs, 'Patenting and Licensing in Genetic Testing: Ethical, Legal, and Social Issues' (2008) 16 Suppl 1 Eur J Hum Genet S10; S Ayme, G Matthijs and S Soini, 'Patenting and Licensing in Genetic Testing' (2008) 16 European Journal of Human Genetics 405; Geertrui van Overwalle and others, 'Models for Facilitating Access to Patents on Genetic Inventions' (2006) 7 Nature Reviews Genetics 143; Geertrui van Overwalle, *Gene Patents and Collaborative*

patent licensing on agreed terms have gained traction as a means of addressing the Covid-19 pandemic,[38] although their effectiveness remains to be seen.[39] Moreover, tools such as model licences can highlight the importance of the public interest and enhance bargaining power in important respects, and have had some recent publicity as a means of safeguarding the public interest.[40] Such licensing models have important potential to advance the public interest.

Third, non-voluntary licensing mechanisms, such as compulsory licensing and Crown use, are an important means by which the public interest is safeguarded. A compulsory licence compels a patentee to allow an applicant use of their patented invention on reasonable terms.[41] In the UK, Crown use provisions permit any authorised department or person to make, use or sell a patented invention 'for the services of the Crown' without the permission of the patentee, so long as it is in the public interest to do so.[42] The Covid-19 pandemic and the lack of widespread international availability of the vaccine have once again focused discussion around compulsory licensing and Crown use.[43] Yet, the difficulties of compulsory licensing in practice demonstrate the obstacles to their effective deployment to address the problems of access that patents on emerging technologies can engender.[44]

Licensing Models: Patent Pools, Clearinghouses, Open Source Models, and Liability Regimes (Cambridge University Press 2009); B Verbeure and others, 'Patent Pools and Diagnostic Testing' (2006) 24 Trends in Biotechnology 115.

[38] Jorge L Contreras and others, 'Pledging Intellectual Property for COVID-19' (2020) 38 Nature Biotechnology 1146.

[39] Karen Walsh and others, 'Intellectual Property Rights and Access in Crisis' (2021) 52 IIC International Review of Intellectual Property and Competition Law 379.

[40] Aisling McMahon, 'Accounting for Ethical Considerations in the Licensing of Patented Biotechnologies and Health-Related Technologies: A Justification' in Naomi Hawkins (ed), *Patenting Biotechnical Innovation: Eligibility, Ethics and Public Interest* (Edward Elgar Publishing 2022); Sarah E Ali-Khan and E Richard Gold, 'Gene Patents Still Alive and Kicking: Their Impact on Provision of Genetic Testing for Long QT Syndrome in the Canadian Public Health-Care System' (2017) 19 Genetics in Medicine 1253.

[41] The TRIPS agreement imposes limitations on compulsory licensing in Art 31: Walsh and others (n 39).

[42] Section 55 and 56 Patents Act 1977. Similar government use provisions may be found in many jurisdictions. The WIPO IP Policy tracker includes a list of existing legislative measures in all countries. For the application of the Crown use provisions in the UK see *Ipcom GmbH & Co KG v Vodafone Group Plc* [2021] EWCA Civ 205

[43] Aisling McMahon, 'Patents, Access to Health and COVID-19-The Role of Compulsory and Government-Use Licensing in Ireland' (2020) 71 Northern Ireland Legal Quarterly 331; Walsh and others (n 39).

[44] Because they are explored in depth elsewhere, these topics are not examined in detail in this book. For discussion see for example McMahon (n 43); Aisling McMahon, 'Global Equitable Access to Vaccines, Medicines and Diagnostics for COVID-19: The

1.5 Reconciling Patents and the Public Interest

As discussed earlier, patent law is underpinned by the public interest – patents do not exist in the abstract as a good in their own right. As has been particularly evident with the Covid-19 crisis, huge sums of public money have been invested in medical research. It is important for public trust that this investment does not result only in private rights and private gain, but that the public interest is appropriately secured.[45] This need not necessarily involve the exclusion or limitation of patent rights entirely, nor public ownership rights for patented inventions. The patent system includes levers at various stages of the patent life cycle, from grant through to exploitation, which are designed to secure this public interest.[46] Depending on the nature of the challenge to the public interest, there are different mechanisms which can be drawn upon to address that challenge, at all stages. Because not all problems are the same, different solutions may be best placed to address each problem. Moreover, in some cases, a multi-faceted approach is best placed to address the problem.[47]

There is no single magic bullet to ensure that patents serve the public interest. As outlined above, robust attention to patent scope, from patentable subject matter, to a rigorous application of patentability criteria, will ensure that the balance struck by the legislature is upheld. Rigorous application of the existing exceptions and exemptions will also serve to ensure that patents are granted only for inventions deemed appropriate on a societal basis. In this respect attention must be paid to principles such as the rules about construction of exemptions, as the wide or narrow construction of these provisions has important implications for what is or is not patentable, with significant knock-on effects for other potentially patentable inventions.[48]

Once a patent is granted, defences to infringement, rules as to flexibilities around enforcement and licensing provisions can serve to provide avenues to advance the public interest. As biotechnologies advance into fields which are served by standards, there is real potential for the fair, reasonable and

---

Role of Patents as Private Governance' (2021) 47 Journal of Medical Ethics 142; Esther van Zimmeren and Geertrui Van Overwalle, 'A Paper Tiger? Compulsory License Regimes for Public Health in Europe' (2011) 42 IIC 4.

45 T Caulfield and others, 'Trust, Patents and Public Perceptions: The Governance of Controversial Biotechnology Research' (2006) 24 Nature Reviews Biotechnology 1352.

46 Dan L Burk and Mark A Lemley, 'Policy Levers in Patent Law' (2003) 89 Virginia Law Review 1575.

47 See N Hawkins, 'Conclusion: Patents and the Public Interest – a Multi-faceted Approach' in N Hawkins (ed), *Patenting Biotechnical Innovation: Eligibility, Ethics and Public Interest* (Edward Elgar Publishing 2022).

48 Walsh (n 23).

non-discriminatory (FRAND) licensing provisions to operate to protect some of the interests discussed earlier.[49]

The importance of building protection of the public interest into licensing from an early stage cannot be overstated. Where universities and public sector organisations are funded by government or charitable funders, then arguably such protections should be put in place to protect the very public whose tax or charitable funds underpin the invention. The funnelling of public money into excessively narrow private interests in this way can then be halted, and the public interest safeguarded through thoughtful and flexible protections. Although bargaining power is often in the hands of the pharmaceutical company, funders are increasingly aware of the importance of safeguarding the public interest of the innovations arising from the research they fund.[50] Moreover, where large publicly funded research infrastructures exist, they can leverage their bargaining power to safeguard the interests of research participants and the public.[51]

Innovation in biotechnology thus raises a number of concerns regarding the way in which patent law functions, with practical, policy and theoretical challenges. This book examines the ways in which the public interest is safeguarded across the patent life cycle and looks at questions of patent eligibility as well as the social and ethical aspects of these types of patented inventions.

2. THE BOOK STRUCTURE AND CONTENTS

The book arises from an academic meeting held in November 2019, organised by the editor, Naomi Hawkins.[52] At this meeting a number of leading patent law scholars from the UK, Europe and Australia presented their current

---

49 See for example discussion of FRAND in Pila (n 22). The ways in which FRAND principles have the potential to work in this manner in the context of medical innovations are explored in Alison Slade, 'Repositioning the Access to Healthcare Debate: FRAND and the Internet of Medical Things' in Naomi Hawkins (ed), *Patenting Biotechnical Innovation: Eligibility, Ethics and Public Interest* (Edward Elgar Publishing 2022).

50 'Wellcome's Approach to Equitable Access to Healthcare Interventions' https://wellcome.org/what-we-do/our-work/access-healthcare-interventions/wellcomes-approach-equitable-access-healthcare-interventions accessed 28 June 2021; Walsh and others (n 39).

51 Hermann Garden, Naomi Hawkins and David Winickoff, (2021) 'Building and Sustaining Collaborative Platforms in Genomics and Biobanks for Health Innovation', OECD Science, Technology and Industry Policy Papers, No. 102, OECD Publishing, Paris, https://doi.org/10.1787/11d960b7-en.

52 This was funded by the ESRC as part of Dr Hawkins's ESRC Future Research Leaders project grant: The Impact of Patents on Translational Research – Non-Invasive Prenatal Diagnosis in Europe and US (grant code ES/K009575/1).

research, and these chapters represent the work of a number of the key contributors. The book explores the debate around the patenting of biotechnologies through different approaches, including doctrinal and socio-legal studies perspectives. This reflects the approaches and perspectives present at the meeting, through which insightful and in-depth discussions influenced the chapters that appear here. These diverse approaches together uncover different aspects of the debates and provide a rich perspective.

This book therefore uses emerging biotechnologies as a lens to examine the operation of the patent system, and the role of the public interest, across the full life cycle of a patent. The book first explores questions of the boundaries of the patent systems – what falls within, or outside of, the system – with examination of the problematic questions which arise at the boundaries. It then goes on to consider the impact of patents on society, addressing key ethical and moral questions, as well as how exploitation of patents and principles of licensing can impact on questions of access.

As outlined above, important and intractable questions continue to arise about questions of patent eligibility, and the early chapters in this book explore the implications of emerging biotechnologies for these issues.

In Chapter 2, Andrew Sagar, Kathleen Liddell, Johnathon Liddicoat and Mateo Aboy consider the boundaries of patent eligibility for fundamental natural laws, abstract ideas and other categories in relation to information-based innovations. They focus on precision medicine, and in particular on the patent eligibility of biomarkers and other nature-based products as well as naturally occurring correlations for diagnosis. These areas have been the focus of developments in the law in recent years, and the authors' examination of the 'legal churn' in the patent law of Europe and in the US draws on the Nuffield Council on Bioethics' report on emerging biotechnology's three characteristics of (i) technological uncertainty, (ii) ambiguity, and (iii) transformative potential as an analytical lens. The authors note the ongoing divergence between jurisdictions, with consequences for the development of innovative biomedical technologies internationally. However, they argue that the legal flux observed in relation to emerging biotechnologies is a result of the difficulties that typically surround emerging technologies, rather than a failure of policymaking per se. In this respect, they argue that empirical evidence can reveal the impact and scope of uncertainty, ambiguity and disruptive potential on the legal system. This evidence can serve to assist policymakers in predicting the interaction of technology with law to mitigate uncertainty. It may also help in the interrogation of the role of patent law in the regulation of emerging technologies, as well as the exploration of the interaction between the law and business models, including the appropriate calibration of the patent system in response to new markets for those emerging technologies.

Chapter 3 continues in examining questions of patent eligibility, focusing particularly on international divergence in the tests for subject matter eligibility. Dianne Nicol and Jane Nielsen explore the eligibility issues that have come into particular focus around the emerging technology of NIPT The validity and scope of patents for NIPT technology are vital for the commercial success of commercial NIPT providers, and have important implications also for public sector providers seeking to offer this testing to their populations. The scope of patent rights is thus implicated in considerations of justice and equity, and exacerbation of health inequalities. Despite the global nature of NIPT technology, divergences in the approach to patent scope internationally have resulted in differences in determinations of the validity of patents for the same invention in different jurisdictions. This chapter explores the nature of these divergences in the key jurisdictions of the USA, Australia and the UK. It outlines the significant case law around patentable subject matter including method of diagnosis and method of treatment claims, and the ways in which that case law has been applied to NIPT inventions. The authors argue that more recent US cases may demonstrate an attempt to find a pathway through the apparent excessive harshness of the *Bilski–Mayo–AMP–Alice* series of Supreme Court cases and to find some certainties for patent holders. Thus, some aspects of NIPT are now being seen as capable of satisfying the subject matter requirement in the USA. In the UK and Australia NIPT has similarly been found to satisfy the requirements of patentable subject matter, although in some cases limitations have been argued in relation to the patentability criteria of insufficiency and inventive step. The authors argue that these recent developments may represent moves to more convergence in the approach to patentable subject matter globally, although recognising the absence of complete uniformity or harmonisation across jurisdictions. Moreover, they argue the analysis of NIPT cases illustrates the resilience within the patent system brought about through judicial interpretation, and how seismic technological shifts can result in courts redrawing or reframing rules in a way that lacks nuance and detail, but which can be reinterpreted and applied by subsequent courts to bring important balance.

In Chapter 4, the discussion around patent eligibility shifts to consider the patentability of products of biofabrication in the context of a 3D biofabricated human organ. Pheh Hoon Lim and Phoebe Li outline the challenges to the patent landscape represented by the creation of a 3D biofabricated organ from human samples of tissue, which offers hope for customised patient treatment such as organ replacement, cell regeneration and access to affordable health care. As the biological components of such biofabricated organ products are already in existence in nature, these technologies raise questions in relation to the blurring of the distinction between discoveries and inventions. The authors query whether a 3D biofabricated human organ (a product created by the bioprinting and multilayering of naturally occurring cells) is a man-made invention, whether it is instead a non-patentable product of nature, despite the need for human intervention to create it. Their considerations draw on parallels between the bioproducts in the Oncomouse and Dolly the sheep cases, and the debate on 'isolated genes' in the Myriad saga. They consider the important jurisdictional divergence in relation to the law for patenting of bioproducts, such that there might not be a single coherent answer internationally to the question of the patent eligibility of bioproducts. Their discussion concludes with key considerations of the public interest: the important issues at stake in relation to biofabrication of human organs, and the need to balance the incentives that patentability might offer as against the risks of monopoly rights for the production of human organs and the related concerns about the ownership of human body parts.

Questions around exceptions to patentability for inventions contrary to *ordre public* and morality are explored by Ella O'Sullivan in Chapter 5. She argues that the approach to interpretation of the exception to patentability for inventions contrary to *ordre public* and morality has persistently marginalised its role in patent law, and that it is this problematic approach to interpretation, rather than the existence of the provision itself, that presents a barrier to the patentability of emerging technologies in this field. The problematic interpretation outlined in the chapter has occurred through incoherent and inconsistent approaches to *ordre public* and morality, resulting in confusion for patent applicants, inventors and potential opponents. In this chapter, O'Sullivan argues that the framework of European and EU fundamental rights which underpins the European patent system requires that the *ordre public* and morality exclusion has serious content and is effectively implemented to address the ethical issues arising from emerging technologies. However, this will only work if the boards of the EPO and the EU adopt a consistent and coherent approach to the exclusion, in contrast to the current approach. The author argues that this current approach has failed to provide a conducive environment for new and emerging technologies, and has instead resulted in a proliferation of inconsistent standards and tests which creates confusion for participants in the patent system, as well as failing to serve its broader social goals. O'Sullivan proposes that a more active engagement with the provision by the EPO would result in a more coherent and appropriate application of the morality provisions, in turn resulting in a more robust European patenting regime.

In Chapter 6, Karen Walsh explores how emerging inventions and the ways in which the scope of exceptions is drawn can disrupt the balance between patent holders and the public. Taking as her starting point the balance between private rights and access on which the European patent system is founded, she considers the way in which the balance is often in practice shifted in favour of rights-holders, and away from the public. She argues that the public interest in favour of access has been shifted too far in favour of expansive patent protection by narrow interpretations of exceptions and exclusions to patentability. Such narrow interpretations are implemented without sufficient consideration of the public interest as a result of the inherent nature of the public interest in patent requirements, with no scope for more explicit examination of relevant public interest factors. She advances this argument by reference to the interpretation of the exception to patentability relating to essentially biological processes by the Boards of Appeal of the European Patent Office (EPO). The line of cases around the interpretation of this exception demonstrates a field of technology where the balance was tilted in favour of patentees, and subsequently re-balanced by later case law. However, this rare reversal is surrounded by controversy. Walsh argues that, because of patent offices', tribunals' and courts' reluctance to further consider the public interest, patentability has been extended into areas previously thought to be against the public interest. As a result, the role of the public interest in patent law is being diluted and its presence within the patent balance is being diminished. With emerging technologies, consideration of the public interest is particularly important in order to avoid overly broad patents and excessive numbers of patents being granted in areas that may not yet be fully developed, and there is need for a clear mechanism by which the public interest can be protected throughout the life cycle of a patent.

Beyond the validity and scope of the patent itself, significant questions arise as to the operation of the patented invention in society: about the invention itself, or about the way in which the patent is exploited. These questions are considered in later chapters of the book.

In Chapter 7, Alison Slade explores the ways in which the obligations of licensing standard essential patents on FRAND terms in the context of the internet of medical things might operate, to ensure the protection of the public interest. The internet of medical things is beginning to revolutionise healthcare provision, empowering individuals to take greater control over monitoring and improving their healthcare in the context of diseases such as diabetes, and increasingly in areas referred to in other chapters of this book, such as NIPT. As in other technological fields relating to the internet of things, however, standard essential patents may represent an obstacle to the development of new technologies as well as access to healthcare treatments. Slade argues that affordable licensing of formal technical standards will help to alleviate this tension between patents and access and reduce barriers to global dissemination of valuable healthcare devices. She argues that in this respect there is scope within the private law contractual FRAND obligation for consideration of the public health objectives. Although contract law limits the scope of consideration to the terms of the agreement, the incompleteness of the FRAND obligation provides sufficient scope for wider public health concerns to be given effect. In contrast, competition law application of FRAND principles is less well suited to the public interest governance role, due to the non-interventionist approach adopted by EU competition authorities to FRAND obligations and an absence of considered engagement with the relevant harm arising in the FRAND context. Although these existing legal frameworks offer some indirect solutions to the protection of public health in the context of the internet of medical things, there is a great deal of uncertainty, and the risk of costly litigation delaying and obstructing innovation is high. Accordingly, Slade argues that greater regulatory oversight of standards in the context of the internet of medical things is important to ensure that the public interest is adequately protected.

In Chapter 8, Aisling McMahon argues for greater regard to the consideration of ethical issues at the patent licensing stage. This chapter focuses on patented biotechnologies and health technologies related to the human body because the rationale for considering ethical issues in the licensing of such patents is heightened due to the potential impacts for human health and life. First, she explores the potential for licensing to be harnessed and used in a manner which complements other regulatory approaches to emerging technologies by discouraging ethically contentious uses of patented technologies via ethical licensing clauses. Second, she argues that ethical use restrictions should be employed in patent licensing, to limit the potential negative repercussions of the way in which patent rights can be used to hinder access to technologies. The chapter concludes by arguing for a deeper, interdisciplinary examination of the control by rights-holders of patented technologies. Such examination should involve both greater scrutiny of the potential for patents to operate as regulatory tools to ensure ethical uses of patented technologies and, the need for oversight over rights-holders' control, to ensure ethical access to downstream patented technologies. This attention to 'ethical licensing' is important if society is to realise the benefits of emerging technologies.

Together, the chapters of this book explore the questions of patentability across the life cycle of the patent, and make an important contribution to patent law scholarship. The book aims to open up scrutiny of the ways in which the public interest can be safeguarded to rebalance the public–private divide. There are multiple ways in which public interest concerns may arise in relation to patented technologies, and patent law and policy therefore must address such concerns as and when they arise, by taking a multi-faceted approach.

2. Patent eligibility and precision medicine: emerging technologies, legal flux and empirical methods

Andrew Sagar, Kathleen Liddell, Johnathon Liddicoat and Mateo Aboy[1]

1. INTRODUCTION

To support emerging technologies, the interests of the innovators of today must be carefully balanced with the interests of those of tomorrow. This constitutes one of the most fundamental challenges of the patent system. The patent system is designed to be applied to new forms of technology. Yet, the uncertainties, ambiguities and transformative potential inherent to an emerging field such as precision medicine present policymakers with difficulties finding the right balance, sometimes leading to unpredictable legal flux.[2]

Information-based innovations, particularly those in the life sciences, often lie close to the boundaries of patent ineligibility proposed for fundamental natural laws, abstract ideas and other categories that could broadly pre-empt future innovation.[3] In the emerging field of precision medicine, the challenges that arise from these boundaries have resulted in considerable change in the patent law of Europe (in the 1990s) and in the US (first in the 1980s and then in the 2010s).[4] This chapter will offer evidence of this 'legal churn' and will argue that it can be better understood and predicted with three characteristics

[1] This work was supported in part by a Novo Nordisk Foundation grant for a scientifically independent Collaborative Research Programme in Biomedical Innovation Law (grant agreement number NNF17SA0027784).

[2] Nuffield Council on Bioethics, 'The threefold challenge of emerging biotechnology' in Nuffield Council on Bioethics, *Emerging Biotechnologies: Technology, Choice and the Public Good* (2012).

[3] MA Heller and RS Eisenberg, 'Can patents deter innovation? The anticommons in biomedical research' (1998) 280 *Science* 698–701.

[4] See D Nicol et al, 'International divergence in gene patenting' (2019) 20 *Annual Review of Genomics and Human Genetics* 519–41.

identified by the Nuffield Council on Bioethics' report on emerging biotechnology: (i) technological uncertainty, (ii) ambiguity and (iii) transformative potential.

This chapter will also discuss the role of empirical studies in mitigating the problematic aspects of legal flux in the context of emerging biotechnology. In particular, it will show that empirical analysis is a useful tool for helping to steer the course through the confusion entailed by tumultuous legal change and can help mitigate some of its more pernicious aspects.

2. PRECISION MEDICINE AND ITS RELATIONSHIP WITH PATENT LAW

Precision medicine has been an 'emerging' field for a long time, with patents on DNA sequences in the US going back to the 1980s.[5] The exact definition of precision medicine is subject to considerable debate. Sometimes the term is used interchangeably with *personalised* or *stratified* medicine, though these can be used in subtly different ways.[6] To offer a definition for the purposes of this chapter, precision medicine aims to revolutionise healthcare by stratifying patients based on personal variation in factors including the patient's genetic information, biochemistry, environment and lifestyle.[7] For example, genomic information can be utilised to identify a patient's vulnerability to a specific disease and the potential efficacy and safety of a treatment. Two facets of precision medicine in particular will be explored in this chapter: (i) biomarkers and other nature-based products; and (ii) naturally occurring correlations for diagnosis.

Biomarkers are useful for tailoring medicine to individuals because they are naturally occurring indicators of biological processes, such as pathogenic processes, or responses to an exposure or intervention, including therapeutic interventions.[8] Examples include genetic sequences, hormone levels, blood pressure and imaging results. Comparing biomarkers with baseline measurements allows the observer to translate the biomarker into a diagnostic or prognostic tool or to identify suitable treatment options. Genomic data, for

---

[5] DB Resnik, 'DNA patents and scientific discovery and innovation: assessing benefits and risks' (2001) 7 *Science and Engineering Ethics* 29–62.

[6] S Day, RC Coombes, L McGrath-Lone, C Schoenborn and H Ward, 'Stratified, precision or personalised medicine? Cancer services in the "real world" of a London hospital' (2017) 39 *Sociology of Health & Illness* 143–58.

[7] EA Ashley, 'Towards precision medicine' (2016) 17 *Nature Reviews Genetics* 507–22.

[8] FDA-NIH Biomarker Working Group, *BEST (Biomarkers, EndpointS, and other Tools)* (2016).

example, can indicate predispositions to particular diseases, predict the efficacy and safety of different treatment options or help develop drugs that are targeted to the causes of diseases rather than their clinical manifestations.

Naturally occurring correlations and relationships also play a central role in precision medicine. The idea is to find correlations between biomarkers – maybe groups of biomarkers – and medical conditions. Once the natural correlation is understood, a patient's health, present and future can be analysed via their biomarkers, for instance, physiological or behavioural data collected via devices such as smartphones, wearables or implantables to predict a patient's health outcomes. As an example, research might involve the use of body-worn sensors to measure gait and balance, facilitating the development of biomarkers that can be employed to assess neurological dysfunction in Parkinson's disease and other movement disorders.[9]

By better tailoring treatment to the needs of patients, precision medicine offers a host of possible benefits. These include radical improvements in the prediction, diagnosis and treatment of common complex diseases and rare diseases.[10] With swifter, more accurate and precise action, medical solutions could be delivered with fewer unnecessary side-effects and interactions with the healthcare system, improving the patient's experience and lowering healthcare costs.[11]

However, precision medicine should not necessarily be viewed as a panacea, nor should the potential costs be obscured. Precision medicine provides probability-based information more often than it does silver bullets. As such, these techniques may be hampered by epistemic difficulties, including the reference class problem, wherein a physician must select a certain class when calculating the likelihood of success of using a treatment for a specific patient.[12] Also, the value of probability-based treatments is limited by the strength of the studies from which it is produced. For example, many studies are currently based primarily on white Anglo-European people, limiting their utility for people of other ethnicities.[13] Furthermore, treating patients with the

---

9 M Aboy, 'How does emerging patent case law in the US and Europe affect precision medicine' (2019) 37 *Nature Biotechnology*, 1118–26.

10 SJ Aronson and HL Rehm, 'Building the foundation for genomics in precision medicine' (2015) 526 *Nature* 336–42.

11 I Akhmetov and RV Bubnov, 'Assessing value of innovative molecular diagnostic tests in the concept of predictive, preventive, and personalized medicine' (2015) 6 *EPMA Journal* 19.

12 D Rowbottom, 'On Probabilities in Personalised Medicine: "The Problem of Untestable Treatments"' in J Liddicoat, J Skopek, and K Liddell, *Precision Medicine: Legal and Ethical Challenges Conference Report* (University of Cambridge 2017) 7.

13 See generally G Sirugo, S Williams and S Tishkoff, 'The missing diversity in human genetic studies' (2019) 177(1) *Cell* 26.

techniques of precision medicine requires doctors and scientists to have access to vast amounts of data, such as population data on genomics and metabolomics.[14] Generating this data is relatively expensive. Identifying correlations also demands significant investment in research and development (R&D), to recognise biomedical inputs, outputs and interactions.

Given these challenges, the commercial viability of precision medicine within a market economy is largely dependent on the intellectual property framework, in particular the patent system for inventions. The patent system is often understood as a structure of rewards and incentives aimed at correcting market failure by encouraging investment into research that would otherwise be vulnerable to free-riding by competitors.[15] In this sense, the patent system is a regulating device that tries to influence human behaviour so that the technology develops, preferably, in a way that delivers its benefits and not its social disbenefits.[16] As will be explained further in section 4.1, however, the patent system needs to be carefully calibrated in such a way as to fulfil this incentivising function while creating space for other innovators – including competitive ones – to adapt and experiment with the initial invention, furthering technological progress.

Creating a regulatory framework, including intellectual property incentives, that harnesses the socially beneficial aspects of precision medicine presents difficulties for regulators and policymakers. It is never easy to identify sweet spots, nor to draw lines when regulating an emerging technology. It is particularly difficult to anticipate how human behaviour in the field of an emerging technology will change in response to legal developments. Section 3 illustrates this in the context of precision medicine and patent protection, drawing attention to the flux that has accompanied these challenges and some of the unexpected impacts brought about by legal change.

---

[14] R Zimmern, 'Health Systems and the Future of Personalised Medicine: A Population Health Perspective' in J Liddicoat, J Skopek and K Liddell, *Precision Medicine: Legal and Ethical Challenges Conference Report* (University of Cambridge 2017) 2.

[15] K Dam, 'The economic underpinnings of patent law' (1994) 23 *Journal of Legal Studies* 247, 249.

[16] R Brownsword and K Yeung, 'Regulating Technologies: Tools, Targets and Thematics' in R Brownsword and K Yeung (eds) *Regulating Technologies* (art Publishing 2008) 3–22; J Black, 'Critical Reflections on Regulation' (2002) 4(4) Centre for Analysis of Risk and Regulation Discussion Paper Series.

3. LEGAL FLUX IN THE PATENT ELIGIBILITY LAWS OF THE US AND EUROPE

3.1 Legal Change in the US

US law on patent eligibility is governed by 35 U.S.C §101, which provides that 'whoever invents or discovers any new and useful process, machine, manufacture, or composition of matter [...] may obtain a patent therefore, subject to the conditions and requirements of this title'. The statute does not elaborate on the meaning of these categories, but judicial case law has evolved interpretation, and the provision operates in tandem with other conditions of patentability including novelty (§102), non-obviousness (§103) and sufficiency of disclosure (§112).

In the 1980 decision *Diamond v Chakrabarty*,[17] the US Supreme Court held that patent rights could be granted pursuant to 35 U.S.C §101 to genetically modified living organisms made by scientists. The decision concerned a bacterium engineered to breakdown crude oil in oil spills, and the Court found it patentable despite its natural origins on the grounds that it was 'markedly different' from what was found in nature. Soon after the decision, a patent over a multicellular organism (an oyster) was judicially recognised,[18] and in 1988 the US Patent and Trademark Office (USPTO) granted a patent over a complex multicellular organism (the Harvard mouse).[19] Thus, the feasibility of patenting products derived from nature was established.

After *Diamond v Chakrabarty*, the USPTO began granting patents with claims directed to isolated, naturally occurring genomic DNA (gDNA). The reasoning was that 'isolated' DNA was not a 'product of nature' because, through advanced human technological intervention, isolated DNA had been cleaved (that is, covalent bonds in its backbone had been chemically severed) or synthesised to consist of just a fraction of a naturally occurring DNA molecule. Thus, for years, both purified and isolated naturally occurring biological substances (as well as the methods to use these biomarkers or other biological substances in methods of detection, diagnosis and treatment of disease) benefited from patent protection at the USPTO.

The US position significantly developed in the 2012 *Mayo v Prometheus* decision.[20] The US Supreme Court held that the defendant's patents were directed to laws of nature and, therefore, were not patentable subject matter.

17 447 U.S. 303 (1980).

18 *In re Allen*, 846 F.2d 77 (Fed. Cir. 1988).

19 US Patent no 4,736,866.

20 *Mayo Collaborative Services v Prometheus Laboratories Inc* 566 US 66 (2012).

Prometheus's patent claimed a process of adjusting a patient's dose of medicine, based on the correlation between metabolite concentrations in blood and dosages of a thiopurine drug that were effective, ineffective or harmful. The patent revoked by the Court had similarities to patents that claimed methods of diagnosis, based on the correlation between genetic variations and disease risks. The case thus raised doubts about the validity of genetic diagnostic claims previously granted by the USPTO.

The Court in *Mayo* held that 'if a law of nature is not patentable, then neither is a process reciting a law of nature, unless that process has additional features that provide practical assurance that the process is more than a drafting effort designed to monopolize the law of nature itself'.[21] The later Supreme Court decision *Alice v CLS Bank International*[22] distilled the reasoning in this case into a two-step test: (1) is the claim directed to a patent ineligible concept (a law of nature)? and, if so, (2) do any additional features of the claim (individually or in combination) transform it into eligible subject matter by setting forth significantly more than the ineligible subject matter?

Although *Alice* was decided two years after *Mayo*, the Court effectively followed these steps in *Mayo*, holding that the dosage-optimising claims triggered the first step, because the correlation 'is a consequence of the ways in which thiopurine compounds are metabolised by the body-entirely natural processes'.[23] The Court then considered whether the claims contained 'significantly more'. The claims involved steps of administering the drug and determining metabolite levels, but the Court found these steps were 'not sufficient to transform unpatentable natural correlations' into patent-eligible claims because they were 'well-understood, routine, conventional activities previously engaged in by scientists who work the field'.[24]

Empirical analyses show the impact *Mayo* had on patents for molecular biology and microbiology. In the six years after the decision, the number of patent applications receiving a §101 rejection rose from 10.5 per cent to 55 per cent, significantly increasing the time and expense involved in prosecuting precision medicine patents.[25] In approximately 35 per cent of cases, applicants were able to overcome these rejections,[26] but these results still show there was a great deal of uncertainty regarding patent eligibility.

---

21 Ibid 8.

22 *Alice Corp v CLS Bank International* 573 US 208 (2014).

23 *Mayo* (n 20) 8.

24 Ibid 10.

25 M Aboy, C Crispo, K Liddell, T Minssen and J Liddicoat, '*Mayo*'s impact on patent applications related to biotechnology, diagnostics and personalized medicine' (2019) 37 *Nature Biotechnology* 513–19.

26 Ibid.

The two-step test developed in *Mayo* and *Alice* did not draw a clear line between patent-ineligible laws of nature and applications of those laws that would be eligible. The test involves flexible principles rather than precise rules, with additional uncertainty arising from Justice Stephen Breyer drawing an analogy between generally applicable natural laws, such as Newton's law of gravity and Einstein's law of relativity, and the drug-specific development of an optimal dosage regime based on individual drug or metabolite relationships.[27] Official responses to this uncertainty have been controversial, especially as the USPTO provided very few concrete examples explaining when a claim directed to a nucleic acid will exhibit 'markedly different' characteristics as compared with naturally occurring nucleic acids.[28]

A year after *Mayo*, the USPTO's decades long practice of granting gDNA patents was finally considered by the US Supreme Court, and with that came further change. In *Association for Molecular Pathology v Myriad*,[29] the US Supreme Court held in a unanimous opinion that isolated, naturally occurring genomic DNA sequences coding for the BRCA1 and BRCA2 genes were *not* patent eligible. The Court applied the 'markedly different' test from *Chakrabarty* and reasoned that 'a naturally occurring DNA segment is a product of nature and not patent eligible merely because it has been isolated'. The Court stated that 'separating a gene from its surrounding genetic material' was 'not an act of invention' since the identification of the BRCA1 and BRCA2 sequences did not involve the modification of the gene itself. Thus, the USPTO practice of granting claims to isolated, naturally occurring genomic DNA (gDNA) sequences came to end. However, it is important to note the Court did find other claims valid. The claims to cDNA were allowed because the Court held that cDNA is not naturally occurring but newly created by scientists.[30] Scientists build a cDNA sequence from mRNA, which results in an exons-only molecule. It differs from naturally occurring DNA which has introns in long sections of DNA as well as exons.[31]

The *Myriad* case was brought by the American Civil Liberties Union (ACLU), which deplored the idea that human DNA sequences could be pri-

---

27 *Mayo* (n 20) 1–2, 8–9, 20.

28 M Aboy, J Liddicoat, K Liddell, M Jordan and C Crespo, 'After *Myriad*, what makes a gene patent claim "markedly different" from nature?' (2017) 35(9) *Nature Biotechnology* 820–5.

29 *Association for Molecular Pathology v Myriad Genetics Inc* 569 US 576 (2013).

30 Ibid 16–17.

31 The Court set out an exception to the ineligibility rule for claims to short strands of cDNA. Although made by scientists, they cannot be claimed if they are indistinguishable from naturally occurring short DNA sequences (which, due to their short length, also lack introns). Ibid.

vately owned in the form of patents.[32] In particular, they opposed the possibility that the corporate owners of patents could prevent people, mostly women, from obtaining more affordable breast cancer risk screening.[33] The ACLU hoped that challenging the patent eligibility of human DNA sequence would put an end to this situation.

As explained, the ACLU won its legal case, in part. It also met some of its wider objectives. Reportedly, it became easier to obtain alternative predictive breast cancer genetic tests from other commercial providers, with the market rapidly expanding in the wake of *Myriad* [34] and, as expected, the numbers of claims to isolated gDNA sequences fell to practically zero.[35] But there are five twists in the tale.

First, an empirical study has shown that the numbers of isolated gDNA sequence patents were already falling prior to the decision, arguably calling into question whether the legal shift that occurred was significant in practice.[36] The rate of decline appears the same even after the *Myriad* decision. Therefore, *Myriad* was probably decided too late to have a major impact; the legal principles of novelty and non-obviousness were already threshing out gDNA patent claims by the time it was decided.

Second, although patents over isolated gDNA plummeted, the numbers of 'gene-*related*' patents continued to increase after *Myriad* because there were various ways in which they could pass the 'markedly different' test. The empirical study classified 'gene-related' patent claims as those that included a sequence of DNA or the word 'DNA' (or other synonyms). The increase of gene-related patents indicated the decision had limited impact on the broader field of genetic technology.

Third, the *Myriad* decision affected patents far beyond isolated gDNA. At first, the decision appeared to be a 'narrowly crafted'[37] outcome that was little more than a 'surgical strike'.[38] Indeed, legal scholars at the time expressed the

---

32 B Carlson, 'Gene patents on the line' (2009) 6 *Biotechnology Healthcare* 8–10.

33 Alexandra Young, 'Prelude to "Pigs Fly:" The Early History of the Myriad Case' (Thesis, Duke University, 2014) pp7–11 https://core.ac.uk/download/pdf/37750098.pdf (last accessed 17 May 2021).

34 Concert Genetics, 'The current landscape of genetic testing' (2017) *Concert Genetics Report* 7.

35 M Aboy, K Liddell, J Liddicoat and C Crespo, '*Myriad*'s impact on gene patents' (2016) 34(11) *Nature Biotechnology* 1119–23.

36 Ibid.

37 GA Schor, V Norviel and IG Cohen, *Brief for amicus curiae Eric S. Lander in support of neither party* (2013).

38 RE Gold, R Cook-Deegan and T Bubela, 'AMP v. Myriad: a surgical strike on Blockbuster business models' (2013) 5(192) *Science Translational Medicine* 192ed9.

view that *Myriad* was limited to patent claims directed to DNA.[39] But a study published five years later showed that when *Myriad* was cited by examiners, 85 per cent of their rejections concerned nature-based product claims beyond isolated DNA. In effect, *Myriad* was being applied to a wide range of nature-based compounds and biomarkers, including proteins, RNA and hormones.[40] This seemed to be an unintended consequence, given the Supreme Court's attempts to contain the bounds of its own decision by conceding that cDNA remained patent eligible. Thus, with regard to nature-based products beyond isolated DNA the legal change was significant – far more so than the Court and some commentators anticipated.

Fourth, commentators speculated that patent applicants (with the help of their patent agents) might be able to 'draft around *Myriad*' and obtain patent claims with equal breadth to isolated gDNA claims. However, empirical research has shown that this has not been easy. Even if in principle it is possible, in more than 70 per cent of patent applications in the three years after *Myriad*, applicants responded by cancelling isolated DNA claims or abandoning the application altogether.[41]

Fifth, the impact of *Mayo* and *Myriad* was not necessarily uniform across the industry. Small, 'patent precarious' organisations are those that rely heavily on patents for competitive advantage. Empirical studies have revealed that these organisations have been particularly affected by the *Mayo* and *Myriad* decisions. Empirical research has found that the number of gene-related patents granted to smaller organisations declined after *Myriad*; before that case, approximately a third of applicants were small organisations, and this number dropped to almost zero in the years after *Myriad* was decided.[42] These unexpected impacts from *Myriad* show unintended effects of the Supreme Court's decision in *Myriad*, illustrating both legal churn and the unforeseen impacts of change in patent practice.

3.2 Legal Change in Europe

In spite of the exclusion of discoveries as such since the European Patent Convention 1973, the approach in Europe appears on paper to be generous. Even so, this has only come about after vociferous, decades-long debates about

---

[39] JS Sherkow and HT Greely, 'The history of patenting genetic material' (2015) 49 *Annual Review of Genetics* 161–82.

[40] M Aboy, C Crespo, K Liddell, J Liddicoat and M Jordan, 'Was the *Myriad* decision a "surgical strike" on isolated DNA patents, or does it have wider impacts?' (2018) 36 *Nature Biotechnology* 1146–9.

[41] Ibid.

[42] Aboy et al (n 35).

ownership of DNA in the 1980s and 1990s, and attempts by European Member States (some which continue today) to restrict the patenting of DNA.

Unlike the US, Europe tackled the challenges of patent incentives for emerging biotechnologies with targeted legislation. The legislative history of the 1998 Biotechnology Directive[43] was a long and tortuous process over a ten-year period during which multiple conflicting views struggled against one another. Concerns over consumer access to gene-based healthcare along with ethical challenges that gene patenting would be contrary to morality and *ordre public* significantly delayed the Directive's promulgation. Some in the European Parliament, for instance, argued for the 'non-patentability of nature', leading to the Parliament's rejection of the first draft of the Directive by the European Commission.[44] Nevertheless, lobbying efforts and economic arguments warning of a loss of competitive advantage with the US and Japan – whose patent practices were comparatively more amenable to gene patenting (at the time) – eventually triumphed, leading to Articles 3 and 5 of the final version of the Biotechnology Directive. These provisions stipulate that isolated DNA sequences are eligible inventions for patent protection.[45] The legal reasoning was that eligible inventions are defined by virtue of having 'technical character' and this can be conferred in different ways such as having a 'technical effect' (including use of the DNA sequence in gene therapy or diagnosis), or – according to the Directive – where the DNA sequence is produced by a technical process.[46]

The opposition to the Biotechnology Directive was diverse and extensive, even involving public street protests. It led to a long period of uncertainty and debate in Europe. The European Patent Office (EPO) made the Directive's norms *de facto* applicable to patent validity in European Patent Convention (EPC) member countries, by adding new provisions in the EPC's Implementing Regulations that reflected the content of the Biotechnology Directive.[47] This further solidified the patentability of isolated gene sequences. Despite this, the

---

[43] ER Gold and A Galloch, 'The European Biotech Directive: past as prologue' (2001) 7(3) *European Law Journal* 331–66.

[44] See the Economic and Social Committee's Opinion on the Proposal for a European Parliament and Council Directive on the legal protection of biotechnological inventions (1996).

[45] Directive 1998/44 on the legal protection of biotechnological inventions [1998] OJ L 213/13 art. 5(2).

[46] Directive 1998/44 recitals 20 and 21, arts 3(2) and 5(2). Art 5(2) states: 'An element [...] produced by means of a technical process, including the sequence or partial sequence of a gene, may constitute a patentable invention.'

[47] European Patent Convention, Implementing Regulations to the Convention on the Grant of European Patents, Rule 23(e), inserted by Decision of the EPO Administrative Council of 16.06.1999 which entered into force on 01.09.1999 (OJ EPO

Greens Party continued opposition proceedings in the EPO against the Howard Florey Institute's patent over the DNA sequence for the human protein relaxin.[48] The Netherlands also brought an action in the Court of Justice for the European Union (CJEU) for annulment under Article 173 of the EC Treaty on a range of grounds, including that it lacked legal basis, infringed the principle of subsidiarity and breached fundamental rights to human dignity and integrity.[49] Both the Greens Party and the Netherlands lost their legal proceedings.

Some countries nevertheless found ways to implement the rules in a narrow manner without objection from European regulators, such as France's adoption of Article L 613-2-1 Intellectual Property Code. This transposes the Biotechnology Directive into French law but limits the scope of gene-sequence claims, and therefore when infringement can be enforced, to 'the specific function disclosed concretely in the description'. The scope of protection was also limited in a function-related way by a decision of the CJEU in 2010, *Monsanto v Cefetra*.[50] Meanwhile, throughout the first decade of the twenty-first century, European ethical advisory groups and other expert reports continued to criticise the appropriateness of patent law granting private property rights over isolated gene sequences.[51]

Eventually, in the 2010s, the debates in Europe dissipated. The position governing patents over biomarkers for precision medicine thus remained that set out in the Biotechnology Directive in 1998 and the EPO's implementing

---

1999, 437 ff). https://archive.epo.org/oj/issues/1999/07/p437.html (accessed 17 May 2021).

[48] T 0272/95 (Relaxin/Howard Florey Institute). The date of the decision was 23.10.2002. www.epo.org/law-practice/case-law-appeals/recent/t950272eu2.html (accessed 17 May 2021).

[49] Case C-377/98, *Netherlands v European Parliament and the Council of Europe*.

[50] The decision in *Monsanto Technology LLC v Cefetra BV* (C-428-08) Judgment of the Court (Grand Chamber) of 6 July 2010 defied clear interpretation. On its face, it held that Article 9 of the Biotechnology Directive meant that protection from a patented gene sequence extended to the material in which it was incorporated (such as a soya bean plant) provided it was not incorporated in a human being and the genetic information performs its function (such as coding for a herbicide resistant protein) – but otherwise not (for example, if the DNA sequence was in lifeless soya meal (produced by cooking and grinding the beans of soya bean plants) so it could not perform its function of conferring herbicide resistance.

[51] For example, European Society for Human Genetics, 'Patents and licensing in genetic testing' [2007] *European Journal of Human Genetics* 1–7; WR Cornish, M Llewelyn and M Adcock, *Intellectual Property Rights and Genetics: A Study into the Impact and Management of Intellectual Property Rights within the Healthcare Sector* (2003); Nuffield Council on Bioethics, *The Ethics of Patenting DNA – A Discussion Paper* (2002).

regulations implemented in 1999, subject to requirements for purpose-limited construction in some countries and by the CJEU's *Monsanto* decision.

The legal position can thus be summarised in two key points. First, naturally occurring products that operate as biomarkers are eligible subject matter under the EPC because biological material, including an isolated DNA sequence or a sequence otherwise produced by a technical process, is eligible for patent protection. Technical character can be achieved by either a technical effect, such as the use of the relevant DNA sequence in a gene therapy, or by the technical processes utilised to purify and produce it outside a cell.[52] Following this logic, the EPO held that the isolated DNA claims in Myriad's European BRCA patents were patent eligible in proceedings in the 2000s,[53] accepting the patent eligibility of biomarkers more generally.

Second, relationships and correlations are also patent eligible if they have technical character. This includes claims to in vitro diagnostic methods based on correlations. The European position on this matter is surprising given the exclusion in Article 53(c) EPC, which states that patents shall not be granted for methods of diagnosis on the human body. This exclusion was examined in the decision G1/04 of the Enlarged Board of Appeal (EBA).[54] The EBA took the view that a diagnostic method is 'practised on the human or animal body' where it involves any interaction with a body that is necessary for the diagnostic step to be performed. However, the EBA also stipulated that the exclusion would only be triggered where the performance of every step with a technical character involved such an interaction with a body.[55] Thus, if a diagnostic method claim involves at least one technical step *independent* from the body, the claim falls outside the exclusion in Article 53(c) EPC (that is, it is patentable subject matter). As such, a step involving in vitro diagnosis with body fluids is outside this exclusion because it does not take place 'on' the human body. This means the issue is simply whether the claimed method of *in vitro* diagnosis has a technical character – which includes data collection with medical instruments, and so on.

---

52 BGH Jan. 19, 2016, X ZR 141/13, http://juris.bundesgerichtshof.de/cgibin/rechtsprechung/document.py?Gericht=bgh&Art=en&Datum=Aktuell&Sort=12288&Seite=2&nr=74097&pos=78&anz=630 [https://perma.cc/Z3GL-HUJQ], translated in A Felmy, 'Receptor Tyrosine Kinase' (2018) 49 *International Review of Intellectual Property and Competition Law* 221, 226–7.

53 T0666/05 (Mutation/University of Utah).

54 *Diagnostic methods*, G 1/04 (2006) OJ EPO 334 (EBA).

55 Ibid.

3.3 Global Flux and On-Going Change

The legal changes in patent law in the US and Europe have taken place at different times, at different paces, and with different organisations at the helm. Yet, unlike the law, scientific organisations and markets operate across countries and do not confine themselves to jurisdictional boundaries. Thus they have had to deal with legal variation not only at national/regional levels but also at an international level, as the approaches of the US, Europe and other countries have converged and diverged.

Summarising Sections 3.1 and 3.2, on paper, the European patent eligibility law is significantly more permissive for biomarkers and correlations. Meanwhile, the US approach is more restrictive. For instance, in the US decision of *Ariosa Diagnostics v Sequenom*[56] the court found that a claim over the discovery that paternally inherited foetal DNA is detectable in pregnant women was *invalid*. In contrast, in *Illumina v Premaitha Health*[57] the UK court held that under the EPC similar claims were *valid* as they incorporated technical steps – namely, creating samples for analysis and detecting nucleic acid.[58]

That said, it should be noted that the degree of divergence between Europe and the US is more complicated than it seems at first glance. The patent system is underpinned by a complex architecture containing multiple regulatory levers for achieving certain policy outcomes, including rules of acquisition and rules of scope of protection.[59] DNA sequences and diagnostic applications of biomarkers could still fall short of other rules of acquisition, for example, by lacking novelty or inventive step, or by the operation of the public morality provision. As such, while a patent claim over an isolated gene may be patent eligible in Europe, it may fail those other steps, and so it should not be assumed that the European approach is 'more generous' than the US when considered from a more systemic perspective.

Although nuanced, the degree of divergence at the international level has implications. An interview-based study has shown that several US-based organisations found the ensuing legal uncertainty problematic and as a result

---

[56] *Ariosa Diagnostics v Sequenom*, 788 F.3d 1371 (2015).

[57] *Illumina Inc v Premaitha Health PLC* [2017] EWHC (Pat) 2930.

[58] Ibid [184]–[189]; the divergent approaches are explored further in Dianne Nicol and Jane Nielsen, 'Non-invasive Prenatal Testing and the Resilience of the Patent System' in N Hawkins (ed), *Patenting Biotechnical Innovation: Eligibility, Ethics and Public Interest* (Edward Elgar Publishing 2022).

[59] K Liddell and J Liddicoat, 'Should we change EU law to disallow DNA patents?' (2016) 7–8 https://ssrn.com/abstract=3102238.

halted development of tests.[60] Even though both groups must meet the §101 requirements to be eligible for US patent protection, the study found only US-headquartered organisations were significantly affected, whereas a number of European entities thought that they could manage the fallout of *Myriad* and *Mayo*.

There were several reasons for the view that US-headquartered organisations were more disadvantaged than European ones. First, US organisations were found to concentrate mainly on the US market where patent protection was weaker, whereas European organisations focused on both US and European markets roughly evenly (some European organisations even focused only on territories outside the US).[61] Second, European organisations made use of business strategies that mitigated the weakness of their US patents. For example, they relied on non-US patent rights, they turned to European and other non-US venture capitalists, they only partnered with European and other non-US organisations and they launched diagnostics tests in Europe before launching in the US using a stronger brand and better data generated from Europe.

Although the legal developments of the 1990s and 2010s are now becoming clearer, organisations and policy makers must still face the situation that legal flux does not appear to be coming to an end. In the US, post-*Mayo/Alice* case law is developing. The US decision *Vanda v West-Ward Pharmaceutical*[62] concerned a method of treatment of schizophrenia with iloperidone, where the drug dosage would be adjusted according to the patient's genotype to reduce the risk of side effects. The court held that the claim was not 'directed to' a natural correlation, but rather an *application* of that correlation. The difference between the claim in *Vanda* and the one in *Mayo* was that the claim in *Mayo* focused on a diagnostic test and involved a dosage adjustment as part of the diagnosis, whereas the claim in *Vanda* incorporated an additional 'administration step' of adjusting the dosage to treat a particular disease. While this may seem like only a slight shift, the less restrictive interpretation of *Mayo* offered by *Vanda* may be a significant evolution of the law on patent eligibility. The case established a relatively easy path to obtain method of *treatment* claims under 35 USC§101; if the *Mayo/Alice* test is managed at Step A, it circumvents the need for an analysis under Step B. In principle, such a development potentially opens a route for patent attorneys to avoid *Mayo* rejections by

---

[60] J Liddicoat, K Liddell and M Aboy, 'The effects of *Myriad* and *Mayo* on molecular test development in the US and Europe: interviews from the frontline' (2019) 22 *Vanderbilt Journal of Entertainment & Technology Law* 785, 824–7.

[61] Ibid.

[62] *Vanda Pharmaceuticals Inc v West-Ward Pharmaceuticals International Ltd* F.3d Vol. 887 1117 (United States Court of Appeals for the Federal Circuit 2018).

transforming diagnostic patent claims into method of treatment claims. Indeed, an early look at the impact of *Vanda* showed that applicants are doing just that: after receiving a *Mayo*-based rejection, patent applicants are transforming their diagnostic claims into method-of-treatment claims to overcome the rejection.[63]

Meanwhile, although the law in Europe continues to be more stable than that in the US, in recent years the *practices* adopted by patentees and potential infringers have shifted. One study has found that 26 out of 169 European clinical molecular laboratories decided not to offer a genetic test to patients due to patent protection.[64] While this may seem to be a relatively low number, this was an increase from 7 per cent in 2008 to 15 per cent in 2017, suggesting gene-related patents are posing an increased burden on Europe laboratories. The non-profit sector was particularly affected, up from 4 per cent in 2008 to 14 per cent in 2017. Almost half the laboratories that chose not to offer clinical genetic tests due to the risk of patent infringement did so without infringement having been previously alleged, indicating that European laboratories have become more aware of patents and have altered their practices accordingly.

4. RATIONALISING THE LEGAL FLUX

Why has patent protection for the emerging technology of precision medicine seen so much legal change? Is it a failure of policymaking? The Nuffield Council on Bioethics Report on Emerging Biotechnology[65] identifies three characteristics of emerging technology that present policymaking challenges. The first challenge is the uncertainty that surrounds the possible impact of an emerging technology. The second, ambiguity, refers to the possibility of different meanings and values being ascribed to the practices and outcomes that arise from emerging technologies. The third characteristic refers to the transformative potential of new fields of technology that can produce substantial changes in the socio-economic and political environment. This trio of issues can work as a lens to better understand the sources of the legal churn surrounding patent eligibility and the emerging technology of precision medicine in Europe and the US since the 1980s. They lend weight to the view that legal flux is not a failure of policymaking but a result of the difficulties that typically surround emerging technologies.

---

63 M Aboy et al, 'One year after *Vanda*, are diagnostics patents transforming into methods of treatment to overcome *Mayo*-based rejections?' (2020) 38 *Nature* 279–83.

64 J Liddicoat, K Liddell, AH McCarthy, S Hogarth, M Aboy, D Nicol, S Patton, M Hopkins, 'Continental drift? Do European clinical genetic testing laboratories have a patent problem?' (2019) 27(7) *European Journal of Human Genetics* 997–1007.

65 Nuffield Council on Bioethics (n 2).

Although the uncertainty, ambiguity and transformative potential of emerging biotechnology make it largely inevitable that it will be difficult to identify an optimal patent policy, it is still possible to distinguish better legal options from worse.[66] To do so – in other words, to evaluate patent systems – one needs to be more aware of assumptions, stakeholders' framings[67] and differences in normative reasoning, *and to seek high-quality empirical information*. In this section we focus predominantly on the latter, demonstrating how, even though empirical analysis may not answer fully the normative questions posed by regulation, such evidence can be valuable for charting a better course through the legal flux.

By shedding light on the impact and scope of uncertainty, ambiguity and disruptive potential with regard to the legal system, empirical analysis can, first, help policymakers to predict how the law will interact with the technology, mitigating the problematic aspects of its inherent uncertainties. Second, empirical studies can help to challenge, or expose as incorrect, the factual bases that underpin different perspectives on the role of patent law in the regulation of emerging technologies such as precision medicine. Third, by examining the relationship between the law and the business models that emerge because of a technology's transformative potential, empirical analysis can assist the appropriate calibration of the patent system by informing regulators how the law will operate within new market landscapes.

4.1 Uncertainty

Typically with emerging technology, one is faced with not knowing, or being uncertain about, the characteristics and boundaries that constitute the object of regulation.[68] Regulation for an unpredictable, emerging technology is therefore often reactive, and can be likened to attempting to hit a moving target. It is likely that a traditional one-size-fits-all solution may be unworkable and tailored solutions will need to be continually assessed and refined.[69] Unsurprisingly, then, the law may oscillate or appear under-developed while it struggles to adapt.

Precision medicine encounters these difficulties with full force. As mentioned in Section 2, there is even disagreement about the definition of precision medicine itself, with multitudes of definitions having been offered. Is it even truly new and emerging? Many doctors argue that they have always practiced

---

66 Ibid para 3.37.

67 Ibid para 3.26 et seq.

68 Ibid para 3.10-3.12.

69 GN Mandel, 'Regulating Emerging Technologies' (2009) *Temple University Legal Studies Research Paper No. 2009-18* 3.

individualised and personalised medicine (for example, by using a thermometer) – though, arguably, precision medicine in its modern conceptualisation is distinguished by the rate of development of diagnostic and treatment options.[70] Other questions include: will the technology actually work, and will it improve medicine in general, for particular classes, or not at all? There is considerable hype and hope, some breakthroughs, but also a significant gap between the promised claims and clinical gains. Considerably more research is needed to validate the clinical utility of interventions in precision medicine. How much will the research cost and how expensive will the personally tailored technology be? Whether precision medicine will yield net social costs is unclear. Many precision medicine treatments can be substantially more costly than conventional treatments.[71] On the other hand, such treatments could reduce healthcare costs downstream, for example by providing better diagnostic accuracy, earlier intervention and reduced encounters with the healthcare system.[72]

These *technological* uncertainties lead to unpredictable interactions with the legal system, giving rise to legal flux. This was evident in the different legal policy developments in different jurisdictions for patent protection for biomarkers and naturally occurring correlations for diagnosis. Will patents 'work' for precision medicine, or will they facilitate disproportionate pricing and make equitable access difficult? Is precision medicine the type of technology that will improve medicine, meaning it is vitally important that society provides appropriate and effective incentives? In the face of uncertainty, there are multiple reasonable answers to these questions, with varying degrees of hedging.

DNA patents could encourage innovation, as without them there could be insufficient incentive to engage in associated research.[73] Nevertheless, the incentive offered by patent protection is far from simple. Like many – if not all – technical fields, precision medicine is driven by cumulative and mixed innovation. For instance, a doctor basing a patient's treatment on the advances of precision medicine will likely rely on information from numerous biomarkers, which themselves are reliant on advances in DNA sequences, platform technologies, diagnostic correlations (which are becoming more multifacto-

---

70 JL Jameson and DL Longo, 'Precision medicine – personalized, problematic, and promising' (2015) 372(23) *New England Journal of Medicine* 2229–34.

71 T Ferkol and P Quinton, 'Precision medicine: at what price?' (2015) 192 American Journal of Respiratory and Critical Care Medicine 658.

72 SP Gavan, AJ Thompson and K Payne, 'The economic case for precision medicine' (2018) 3 *Expert Review of Precision Medicine and Drug Development* 1.

73 See also K Liddell et al, 'Patents as incentives for translational and evaluative research: the case of genetic tests and their improved clinical performance' (2008) 3 *Intellectual Property Quarterly* 286.

rial) and prescribing practices. Monopolistic intellectual property may permit holders to set high licensing prices that could hinder the diffusion of the relevant technology, producing a co-ordination failure with other innovators and chilling the follow-on investment needed for future inventions based on the initial patent.[74] Accordingly, patent exclusivities arguably need to be limited to avoid a 'tragedy of the anticommons'.[75] However, the tipping point between sufficient incentives and sufficient freedom to use and build on knowledge is uncertain and potentially ever-changing. Consequently this leads to a lot of disagreement and legal 'experimentation' – and thus, the legal flux observed in both the US and Europe.

In *Mayo*, the Court was animated by the concern that too much patenting may impede innovation, maintaining that 'there is a danger that the grant of patents that tie up [the use of natural laws] will inhibit future innovation premised upon them'.[76] However, the Court lacked sufficient evidence about research and investment in the diagnostics industry as well as the role of cumulative innovation in diagnostics to know whether such patents in practice do obstruct innovation by competitors. Subsequent courts thus doubted the extent to which they should follow or distinguish the decision. And in the EU, as earlier mentioned, part of the drive towards harmonisation of the law in this area was to remain competitive in the global research economy when the future of biotechnology remained uncertain.[77] Different countries weighted the uncertain opportunities and risks differently, and reached different views about the benefits and problems of patent protection in this field. Legal change was eventually ushered in with the Biotechnology Directive, but only after extensive debate; it remained highly controversial and only partially harmonised Member States' patent norms.

Empirical legal analysis can be critical in mitigating the instability of legal flux, since it can provide decision-makers with more, albeit inevitably incomplete, information concerning the mutual interaction of the law and emerging technologies. For instance, information about patenting trends and their impact on innovation can enable more rational regulatory experimentation.

To illustrate, one study (described above) noted an increase in the number of European clinical molecular laboratories abstaining from offering genetic tests for fear of patent infringement, choosing instead to respect intellectual

---

[74] See also C Sweet and D Eterovic, 'Do patent rights matter? 40 years of innovation, complexity and productivity?' (2019) 115 *World Development* 78–93; Heller and Eisenberg (n 3).

[75] Heller and Eisenberg (n 3).

[76] *Mayo* (n 20) 17.

[77] Gold and Galloch (n 44).

property rights.[78] In the absence of empirical legal studies, policymakers lack awareness of the operation of the 'law in action', or how the law interacts with other practices that fuel the emergence of precision medicine. With evidence like this, policymakers are better placed to consider issues like patent eligibility. This evidence offers insights as to whether patents do incentivise research and, by extension, facilitate access to diagnostic tests. Such data may be also used to challenge the assumption that patents ubiquitously incentivise research, instead providing evidence that in some cases they may disincentivise competition and cumulative innovation by other laboratories. Alternatively, a more conservative interpretation may be embraced, and a policymaker may instead be encouraged to, for example, provide increased legal support to laboratories facing patent issues.

4.2 Ambiguity

The Nuffield Council's second characteristic for the difficulties of developing policy for an emerging technology is ambiguity. The Council uses the term to describe how people can view the same information.[79] Meanings can be highly varied in a pluralist society where individual values, beliefs and needs are so wide-ranging. Precision medicine and early diagnosis changes the nature of what it means to be ill, which patients have the 'same' disease and prognosis, and which patients are eligible for particular treatments. It changes the relationships between people, and not all people will view this shift favourably, or as favourably. Their views on the benefits of patent protection could be coloured by this, and their view of patent protection could depend on whether it creates better healthcare opportunities for all or only for some, and whether private ownership of nature-related innovation is ethically acceptable. The law develops based on the contributions, participation and influence of different actors that approach ambiguities using different frameworks, and so the evolution of the law becomes a tug-of-war. Deciding how to reconcile these different perspectives on the same phenomena is a challenge for lawmakers. As interests collide and struggle against one another, the law often shifts and changes.

In the US, *Myriad* can be seen as a story of competing values and interpretive frameworks. The ACLU believed genetic tests for breast and ovarian cancers should be less costly and that DNA sequences should not be subject to private control. This view was at odds with that of the innovative company, Myriad, which wanted to maximise the value of its investment through exploitation of

78 Liddicoat et al (n 65).
79 Nuffield Council on Bioethics (n 2) para 3.13.

a broad patent and then plough more money into cancer treatment research. Meanwhile, the Court saw the problem of patenting in this area as pre-empting future innovation by third parties. Accordingly, the impetus for the litigation in the first place was a clash of values, with the ACLU advocating for access to diagnostic tests for patients while Myriad was seeking to maximise return on investment to drive forwards further research. Finally, the judicial decision was based on yet still another set of values, namely a concern for impediments on scientific freedoms.

At the heart of the *Mayo* and *Myriad* decisions is another ambiguity: how actually to characterise innovations in precision medicine. Should we regard isolated biomarkers as new compositions or natural products? Should we regard patents that claim the use of medical correlations as a claim over a natural law or as a method of treatment or diagnosis? Perhaps the difficulties associated with resolving the latter ambiguity may explain *Vanda*. To some extent, *Vanda* was criticised as an artificial gloss on *Mayo/Alice*,[80] but that artificiality could come from the unworkable nature of those earlier decisions and an attempt to step towards a different conceptualisation of patents as incentives for improved medical treatments.

From the European perspective, the Biotechnology Directive also illustrates how ambiguity can drive legal flux. During the ten-year dialogue preceding the law's entry into force, divergent policy concerns led to a Directive that represents a compromise between Parliament, the Commission, and the Council. As discussed in section 3.2, some proponents of the Directive viewed patenting biotechnology as a means of facilitating scientific innovation and keeping European R&D competitive with the US.[81] Others, however, saw it as a system that led to private proprietary domination of common heritage, which in turn led to vulnerable patients being charged excessive prices in their time of need.[82] As a consequence, the Directive's drafting was an arduous, ever-changing process, and this collision of conflicting, possibly incompatible views on biotechnology in the EU can be seen as a direct reflection of the latent ambiguity that characterises all emerging technologies.

Empirical analysis cannot fully engage with normative questions concerning, for example, in-principle ethical prohibitions to gene patenting, but can help to challenge the assumptions from which some normative claims are derived. For instance, one ethical claim might be that it is 'wrong' to patent life. However, patenting trends may reveal that what is being patented is

---

80 M Aboy et al, 'One year after *Vanda*, are diagnostics patents transforming into methods of treatment to overcome *Mayo*-based rejections?' (2020) 38 *Nature* 279–83.

81 Gold and Galloch (n 44).

82 Ibid.

isolated DNA. This, in turn, may suggest the patent is not a 'patent on life', and does not interfere with individual human beings' or organisms' freedoms. These arguments can then be recognised as misplaced.[83]

Similarly, empirical analysis alone cannot reconcile the fundamental frameworks upon which conflicting views are predicated, nor can it decide which is a more desirable approach, but it can challenge the factual bases that give rise to contrasting views on the relevance of patents. For instance, if patents do not actually stimulate economic growth, then the view that patents related to precision medicine are beneficial for economic growth is misguided. But empirical analysis might also show that calls for free use of isolated DNA sequences to revolutionise patient care are also misdirected. Alternatively, if the threat of patent enforcement is found to disincentivise potentially innovative research, then perhaps policymakers might be encouraged to reject arguments proselytising on the virtues of patents as stimulants of social progress. As such, while an inevitable characteristic of emerging biotechnologies is their tendency to cause different views on the same phenomena, empirical legal studies can at least reduce unnecessary divergence in opinions, narrowing down arguments about how the patent system should respond to these ambiguities.

4.3 Transformative Potential

The third challenge identified in the Nuffield Council's report concerns the transformative potential of emerging technologies, or, putting it another way, their capacity to change and disrupt social practices now or in the future. Where mainstream practices begin to ossify and stagnate, radical innovations may perform old functions significantly more efficiently or differently, thus rearranging at least some aspects of the existing political economy.[84] As an example, some have argued that companies like Uber and Lyft have drastically altered the ways in which transportation services are organised – though, it should be noted, others contend that the term 'disruption' has been appropriated by industry players as a cloak for regulatory arbitrage.[85] Similarly, as mentioned earlier, there are some who doubt whether precision medicine is truly as revolutionary as industry players would have us believe. In any case, recent technological advancements in precision medicine have heralded significant changes in the marketplace. This section considers how that disruption – or in

---

83 CM Holman, 'Debunking the myth that whole-genome sequencing infringes thousands of gene patents' (2012) 30(3) *Nature Biotechnology* 240, 241–2; see also, Liddell and Liddicoat (n 48) 3.

84 Nuffield Council on Bioethics (n 2) para 3.22-3.25.

85 J Prassl, *Humans as a Service* (Oxford University Press 2018) 34–7.

other words, transformation – has contributed to the legal flux in subject matter patentability, and the importance of related empirical analyses therein.

The emergence of precision medicine has offered increased opportunities for firms to capture value, and accordingly, recent decades have seen two related trends in the diagnostics industry.[86] The first trend revolves around the 'corporatisation' of R&D, in which there has been increased involvement of private sector actors in the discovery and development of biomarkers. As Hogarth contends, while publicly funded research by academic–clinical collaborations remains relevant, industry's role in the development of diagnostic signatures has become gradually more pronounced.[87]

The second trend is the 'pharmaceuticalisation' of the diagnostics industry, wherein diagnostics firms have been adopting pharmaceutical industry business strategies to capture value, such as through biomarker patenting rather than relying on platform patents (that is, 'on methods, instrumentation, or reaction chemistries').[88] This offers a range of commercialisation options. Some firms use biomarker patents by commercialising their own laboratory-developed tests, whereas others adopt a 'dual approach', selling kits to other laboratories that are protected by biomarker and platform patents. Some firms claim patents over specific individual biomarkers, whereas others patent combinations of markers along with interpretive algorithms.

These dynamics of precision medicine pose difficulties for regulators. The new value propositions in the diagnostics industry, while attractive to entrepreneurs and corporate culture, affected the costs and competitiveness of diagnostics, raising concerns for patients and healthcare payors. The *Mayo* and *Myriad* decisions were a reaction, but disruptive in their own way for the nascent business models. By narrowing the patent protection upon which they had recently become reliant, firms had to adjust the pharmaceuticalisation of the diagnostics industry. The effect was more marked for SMEs which had weak or non-existent patent portfolios. They were left with limited competitive advantages, fewer assets and more dependence on venture capitalist investment (absent the patent protection that often attracts them), and had less resources to spend on legal advice to respond to patent examination rejections.[89] In contrast, larger organisations with more varied patent portfolios withstood the disruptions to proprietary business models entailed by the kind of flux affecting US patent eligibility law in the 2010s.

---

[86] S Hogarth, 'Personalized medicine – a typology' (2016) *Briefing for CADTH* 10–11.

[87] Ibid.

[88] Ibid 10.

[89] Liddicoat et al (n 61) 29–30.

Empirical evidence can be invaluable here, showing how regulation has impacted, and may continue to impact, nascent market practices. For instance, policymakers might consider how *Mayo* and *Myriad* interact with the new political economy of the diagnostics industry, namely the increasing prevalence and business significance of patents for diagnostics companies. By integrating findings like these into future decision-making, policymakers could reduce the risk of further unintended negative consequences of legal flux. Without the relevant information, lawmakers may be embracing policies based on an understanding of markets that no longer exist, but rather have been moderately or even radically reshaped by the influence of emerging technologies.

5. CONCLUSIONS

This chapter has described significant and on-going legal flux in the patent eligibility laws of the US and Europe, with major implications for precision medicine. The three characteristics of emerging technologies identified by the Nuffield Council on Bioethics provide an analytical framework that helps rationalise, at least in large part, the fluctuating development. The uncertain, ambiguous and transformative nature of an emerging field such as precision medicine creates unpredictable challenges for policymakers and can thus trigger a lengthy period of legal flux.

Some legal flux, including in patent eligibility laws, can lead to beneficial results, as the flexibilities entailed by more open-textured norms can allow for a dynamic system that is sensitive to case-specific nuances. In addition, disruption to prevailing legal and business norms can pave the way for radical, creative ideas and approaches that enrich our society. But in other ways, legal flux can be deeply problematic, leading to false assumptions and framing on the part of stakeholders, chilling investments that are deemed too risky and unsettling policymakers who do not know where, how or if to engage with the issues presented by emerging technology. In other words, there is a distinction between legal flux that creates opportunities by offering flexibility and nuance, and legal flux that obscures possibilities by generating confusion and disarray.

Empirical analysis can help distinguish the problematic aspects of a changing social and legal environment. It informs policymakers and stakeholders how particular laws operate in practice. This in turn gives them more information about the degree, nature and trends in the changes, and the pace with which legal uncertainty is being resolved. It also informs stakeholders of what more precisely are the 'rules of the game', and whether and when these are becoming more established. As a result, despite the inherently indeterminate nature of emerging technology, empirical legal studies can help chart a path through the ensuing legal flux and offer more rational bases for decision-making by policymakers, regulators, judges and stakeholders. As this chapter has shown, empirical research about the extent and impact of patent protection relevant to precision medicine is a case in point.

The insights of empirical legal studies remain critical for patent eligibility of DNA sequences, other biomarkers and medically relevant correlations. This is because the uncertainties, ambiguities and disruptions entailed by precision medicine are still ongoing, and are likely to persist as long as precision medicine remains an emerging form of technology. Other areas of emerging technology can also glean important insights from the story of precision medicine. As can be seen from the legal responses in Europe and the US since the 1980s, the emergence of a new field such as precision medicine and its interaction with intellectual property protection may encourage decision-makers to adjust the patent system in various ways. Outcomes are likely to be unpredictable as the system strives to balance the interests of the innovators of today and those of tomorrow.

3. Non-invasive prenatal testing and the resilience of the patent system

Dianne Nicol and Jane Nielsen

1. INTRODUCTION

In 1996 researchers led by Dennis Lo, based at Oxford University in the UK, discovered a new type of DNA floating freely in the blood stream of an expectant mother, which they called cell free foetal DNA (cffDNA).[1] A new form of prenatal diagnostic testing, which later became known as non-invasive prenatal testing (NIPT), utilised this presence of cffDNA in the maternal blood stream. However, it took more than ten years for NIPT to become a mainstream form of testing for foetal abnormalities, requiring significant advances in genomic techniques.[2] Commercial development followed quickly, and in the US alone, four companies were offering NIPT by late 2011.[3] The potential market for NIPT is enormous – essentially, any pregnant woman around the globe.[4] We are already seeing expansion in the availability of NIPT both geographically and in the range of medical indications for which it is offered.[5]

Various alternatives to NIPT already existed in the prior art, including a number of different cytogenic techniques for detecting chromosomal abnormalities, with cells sourced through amniocentesis (from foetal cells in the

---

[1] YM Dennis Lo et al. 'Presence of Fetal DNA in Maternal Plasma and Serum' (1997) *350 Lancet* 485–7.

[2] Rossa WK Chiu and YM Dennis Lo, 'Non-invasive Prenatal Diagnosis by Fetal Nucleic Acid Analysis in Maternal Plasma: the Coming of Age' (2011) 16 *Seminars in International Fetal and Neonatal Medicine* 88–93.

[3] Ashwin Agarwal et al., 'Commercial Landscape of Noninvasive Prenatal Testing in the United States' (2013) 33 *Prenatal Diagnosis* 521.

[4] Vardit Ravitsky, *Non-Invasive Prenatal Testing (NIPT) Identifying Key Clinical, Ethical, Social, Legal and Policy Issues* (London: Nuffield Council on Bioethics Background Paper, 2015), 4–5.

[5] See for example Anthony R Gregg et al., 'Noninvasive Prenatal Screening for Fetal Aneuploidy, 2016 Update: A Position Statement of the American College of Medical Genetics and Genomics' (2016) 18 *Genetics in Medicine* 1056.

amniotic fluid), chorionic villus sampling (from placental cells in the placental wall), cordocentesis (from foetal cells in the umbilical cord) and maternal blood sampling (foetal cells in maternal blood).[6] One of the main problems with these techniques is that the majority are invasive and carry a risk of miscarriage.[7] These concerns about the invasiveness of some of the sampling techniques, together with risks associated with false positives, added to the appeal of a non-invasive technique with greater accuracy and a lower rate of false positives.[8] It is perhaps little wonder, then, that NIPT has been described as the 'vanguard of genomic medicine'.[9]

The commercial landscape for NIPT is growing rapidly in its complexity.[10] The foundational patent arising from the research conducted at Oxford University has been the main source of concern for the developing NIPT industry. This is because most forms of NIPT were believed to fall within its scope.[11] The patent, designated US6258540 in the US (the '540 patent'), was filed widely in many jurisdictions.[12] It was first assigned to Isis Innovation Ltd, the technology transfer arm of Oxford University. Isis Innovation Ltd awarded worldwide exclusive rights to this patent portfolio to Sequenom in 2005 and assigned the patent to Sequenom in 2014. Since then, Sequenom has been acquired by LabCorp and has entered into a patent pooling agreement with Illumina relating to these and a raft of other NIPT patents.[13] As such, patent rights relating to the significant research breakthroughs of Lo and others in the 1990s essentially became controlled by licensees of the assignees of the assignees of the original researchers. Whether or not their patent enforcement practices actually reflect the intentions of the original researchers has become largely irrelevant.

---

[6] R Akolekar et al., 'Procedure-Related Risk of Miscarriage Following Amniocentesis and Chorionic Villus Sampling: A Systematic Review and Meta-Analysis' (2015) 45 *Ultrasound in Obstetrics & Gynecology* 16.

[7] Ibid.

[8] Mollie Minear et al., 'Noninvasive Prenatal Genetic Testing: Current and Emerging Ethical, Legal, and Social Issues' (2015) 16 *Annual Review of Genomics and Human Genetics* 369.

[9] Lisa Hui and Diana W Bianchi, 'Noninvasive Prenatal DNA Testing: The Vanguard of Genomic Medicine' (2017) 68 *Annual Review of Medicine* 459–72.

[10] For an early study of the complexity of the NIPT commercial landscape see Agarwal et al., above n 3.

[11] Naomi Hawkins, Dianne Nicol, Robert Cook-Deegan and Subhashini Chandrasekharan, 'The Continuing Saga of Patents and Non-invasive Prenatal Testing' (2019) 39 *Prenatal Diagnosis* 441–7.

[12] Ibid.

[13] Ibid.

Although this is not uncommon practice, the particularly commercial nature of delivery of NIPT services, together with the importance of these services to women of child-bearing age globally, lends particular piquancy to inquiries into the scope of patent rights and the limits on their enforcement. Justice and equity, and the exacerbation of health inequalities, are already being raised as real risks.[14] Lack of public funding or insurance cover for NIPT is to blame to some extent, but the zones of exclusivity over NIPT caused by exercise of patent rights play no small part.[15]

This chapter explores the patent landscape for NIPT, with particular focus on the ways in which recent judicial decisions on the patentability of DNA sequences and related processes impact on the validity of NIPT patents. Three jurisdictions – the USA, the UK and Australia – have been chosen for this analysis. NIPT patent litigation has been on foot in each of these jurisdictions, making them useful comparators in examining the differing ways in which ostensibly harmonised national patent laws may vary in their application between jurisdictions.

2. BACKDROP

Patents claiming exclusive rights over DNA sequences, proteins, stem cells and other research materials and methods of their use have for many years littered the genomic technology landscape. If they are to fulfil their intended purpose, patents of this nature should facilitate the development of novel therapies and treatments, by encouraging investment in further innovation and product development. But an alternative scenario is that they could have a stifling effect by closing off new areas of research and technological innovation. Many countries have wrestled with the question of whether patents of this nature have had the effect of stifling or promoting innovation, and evidence either way is equivocal at best. Law reform agencies have so far been cautious when considering the need to reform the law, recommending nuanced responses.[16] This reflects the delicate balance that intellectual property laws must maintain in protecting the interests of technology leaders, follow-on innovators and society more broadly.

The primary concern for those who fear the stifling effect of patents has been that whole areas of research and development could be blocked off by the zones of exclusivity created by broad patent claims. The broader the patent

---

14 Ravitsky, above n 4, 14.

15 Ravitsky, above n 4, 16.

16 See, for example, Australian Law Reform Commission, *Genes and Ingenuity: Gene Patenting and Human Health* (Report 99 Commonwealth of Australia, Canberra, 2004), 14.

claims, the more expansive the exclusivity zone would be. In 1998 Michael Heller and Rebecca Eisenberg wrote that an even more serious problem could arise if there were multiple overlapping zones of exclusivity, creating a series of 'tollbooths on the road to product development', their so-called anticommons tragedy.[17]

Despite these concerns, however, it appears that, in most countries where these types of patents did exist, their owners were largely indifferent about policing their exclusivity zones, even going so far as to abandon them entirely.[18] Yet there have been instances of more aggressive enforcement strategies, including insistence on exclusivity or imposition of onerous licensing conditions.[19] There are continued concerns that the environment could be ripe for the emergence of more strategic patent enforcement activities by the for-profit sector, to the detriment of public access to new healthcare innovations. Given the breadth of the Lo patent and the commercial nature of NIPT testing, this area of technology might have provided the paradigmatic testing ground to explore the consequences of strategic patent enforcement. One of the questions to be explored in this chapter is whether recent judicial decisions examining the scope of patentable subject matter may have actually circumvented this eventuality.

The development of new healthcare products is without doubt risky and expensive, and is not ordinarily the province of the non-profit research sector.[20] Rather, healthcare technology development is, at some point in the product development pipeline, routinely handed to for-profit partners, who are better equipped to manage these issues. This usually occurs before expensive phase 3 clinical trials commence.[21] Patents provide them with exclusivity in the marketplace, albeit temporarily (usually for 20 years). There is much that could

---

17 Michael A Heller and Rebecca S Eisenberg, 'Can Patents Deter Innovation? The Anticommons in Biomedical Research' (1998) 280 *Science* 698.

18 Sibylle Gaisser et al., 'The Phantom Menace of Gene Patents' (2009) 458 *Nature* 207; John Liddicoat, Tess Whitton and Dianne Nicol, 'Are the Gene Patent Storm Clouds Dissipating? A Global Snapshot' (2015) 33 *Nature Biotechnology* 347; Naomi Hawkins, 'The Impact of Human Gene Patents on Genetic Testing in the United Kingdom' (2011) 13 *Genetics in Medicine* 320.

19 In the diagnostic testing space, for example, see Jon F Merz et al., 'Diagnostic Testing Fails the Test' (2002) 415 *Nature* 577; Mildred K Cho et al., 'Effect of Patents and Licenses on the Provision of Clinical Genetic Testing Services' (2003) 5 *Journal of Molecular Diagnostics* 3.

20 Dianne Nicol et al., 'International Divergence in Gene Patenting' (2019) 20 *Annual Review of Genomics and Human Genetics* 519.

21 Dianne Nicol et al., *The Innovation Pool in Biotechnology: The Role of Patents in Facilitating Innovation* (Hobart: Centre for Law and Genetics Occasional Paper No. 8, 2014), 78–80.

be said about whether or not the current patent system achieves this purpose. Particular concerns have been raised about the role of patents in genetic diagnostic testing, for example.[22] Unlike drugs, the path to market for genetic diagnostic tests is very much truncated. Once a gene has been isolated and sequenced, and its function deciphered, it does not take long before a genetic diagnostic test is market-ready.

Yet for many years, patent offices appeared comfortable with the idea that chemical isolation of a gene creates subject matter that is patentable, provided that it is accompanied by some sort of evidence of useful function. Likewise, methods using these isolated chemicals could also be the subject of valid patent claims. There are, of course, other requirements for patentability, including novelty, inventive step and full disclosure, which impose restrictions on the scope of the exclusivity zone that can be claimed. But the more fundamental question is whether DNA sequences and methods of using them are actually the types of subject matter that cross the patentability threshold. The highest courts in the US and Australia have now addressed this issue head-on, declaring that isolated DNA sequences that are identical to their natural counterparts are not patentable subject matter.[23]

These US and Australian court decisions both coincidentally relate to patents owned by Myriad Genetics, Inc, which claimed patent rights to DNA sequences known to be connected with higher-than-average propensity to develop breast cancers (the so-called BRCA genes). These patents were particularly controversial because of the way that Myriad Genetics chose to exercise its exclusive rights in the US, and the way that its exclusive licensee threatened to do so in Australia.[24] Because the DNA sequences in issue were connected with increased propensity to develop breast cancer, there was a major public health issue at stake. Both courts were emphatic (though for slightly different reasons – explained further below) that these patent claims to BRCA sequences were invalid because they failed to cross the patentable subject matter threshold.

The US Supreme Court has made other inroads into the perceived problems with patents claiming rights to controversial subject matter. The Court has held

---

[22] Merz et al., above n 19; Cho et al., above n 19; Secretary's Advisory Committee on Genetics, Health and Society ('SACGHS'), *Gene Patents and Licensing Practices and their Impact on Patient Access to Genetic Tests* (Washington DC, 2010).

[23] *Association for Molecular Pathology v Myriad Genetics Inc* 133 S Ct 2107 (2013) ('*AMP v Myriad*'); *D'Arcy v Myriad Genetics Inc* [2015] HCA 35 ('*D'Arcy*').

[24] E Richard Gold and Julia Carbone, 'Myriad Genetics: In the Eye of the Policy Storm' (2010) 12 (Supp 4) *Genetics in Medicine* 532; Dianne Nicol and John Liddicoat, 'Do Patents Impede the Provision of Genetic Tests in Australia?' (2013) 37 *Australian Health Review* 281.

that patents claiming methods of using subject matter derived from the natural or abstract world will be invalid unless the methods themselves are truly innovative (the Australian High Court has not yet had chance to deliberate on this question). The broad consequence of these decisions is that parties claiming rights to technological innovations such as Dolly the sheep[25] and, indeed, NIPT have failed to satisfy the lower US courts that these are patent eligible. As noted by Reyna J in the opinion of the court in the Federal Circuit decision in *Ariosa Diagnostics Inc v Sequenom Inc*, relating to the patent eligibility of the Lo patent: 'We do not disagree that detecting cffDNA in maternal plasma or serum that before was discarded as waste material is a positive and valuable contribution to science. But even such valuable contributions can fall short of statutory patentable subject matter, as it does here.'[26]

Linn J, in concurrence, expressed further concern about the exclusion of this 'meritorious' invention, noting that, but for the Supreme Court decision in *Mayo* (discussed below), he saw 'no reason, in policy or in statute, why this breakthrough invention should be deemed patent ineligible'.[27]

Some commentators have been more sanguine in their evaluation of the BRCA cases, in particular arguing that the patents in issue were claiming far too expansive zones of exclusivity than the innovative concept that they disclosed, and should not be held valid.[28] Still, it is not at all clear that we now have the balance right: it could have shifted too far in favour of technology followers rather than leaders; or alternatively it may still be that clever patent attorneys can still find creative ways around the restrictions imposed by the courts.[29] A third alternative is that the courts themselves are able to find their own ways of setting limits around their decisions.[30] We will see later in this

[25] *In re Roslin Institute (Edinburgh)*, 750 F.3d 1333 (Fed. Cir. 2014).

[26] *Ariosa Diagnostics Inc v Sequenom Inc*, 788 F.3d 1371 (Fed Cir, 2015), page 16 of Reyna J's slip opinion.

[27] *Ariosa Diagnostics Inc v Sequenom Inc*, 788 F.3d 1371 (Fed Cir, 2015), pages 4–5 of Linn J's slip opinion.

[28] See, for example, Richard E Gold, Robert Cook-Deegan and Tania Bubela, '*AMP* v. *Myriad*: A Surgical Strike on Blockbuster Business Models' (2013) 5 *Science Translational Medicine* 192ed9.

[29] Mateo Aboy et al., 'After *Myriad*, What Types of Claim Amendments Change a Patent Ineligible Isolated Gene Claim into an Eligible Patent Claim that Is "Markedly Different" from Nature?' (2018) 35 *Nature Biotechnology* 820; Robert Cook-Deegan et al., 'The New Controversy in Genetic Testing: Clinical Data as Trade Secrets?' (2013) 21 *European Journal of Human Genetics* 585; Rochelle Dreyfuss, Jane Nielsen and Dianne Nicol, 'Patenting Nature – A Comparative Perspective' (2018) 6 *Journal of Law and the Biosciences* 1.

[30] Dreyfuss et al., ibid.

chapter that this last option appears to be finding some traction, particularly in the US, including in the NIPT sphere.

3. THE EARLY NIPT PATENT JURISPRUDENCE

3.1 Key Prior Developments in Patentable Subject Matter Jurisprudence

Although the 1994 Agreement on Trade-related Aspects of Intellectual Property Rights (TRIPs) ostensibly harmonises the intellectual property laws of World Trade Organization member states, it only provides broad parameters on the elements that must be included in national laws.[31] Hence, the requirement in Article 27 of TRIPs that patents must be 'made available for any inventions, whether products or processes, in all fields of technology' does not set parameters on what constitutes a patentable invention. Rather, it is left to member states to determine for themselves the parameters of patentable subject matter. The brief overview presented below of the differing approaches to patentability in recent jurisprudence in the United States (US), Europe and Australia illustrates how much variation there can be.

Despite the same nature-based technologies being patented across multiple jurisdictions, and the validity of those patents being challenged at around the same time, differing outcomes have meant the patentable subject matter landscape across jurisdictions has diverged.[32] This is due in part to slight variations in subject matter actually being claimed in these patents across jurisdictions, but it is rather more as a result of differing approaches to the question of what legally qualifies as patentable subject matter.

3.1.1 Patentable subject matter in the US

Much has been written on recent developments in patentable subject matter jurisprudence under US law, and what follows is not intended to be an exhaustive exposition of US case law. Rather, it seeks to set the scene for the analysis of NIPT patent decisions that follows. US patentable subject matter jurisprudence centres on section 101 of US patent legislation: 'Whoever invents or discovers any new and useful process, machine, manufacture, or composition of matter [...] may obtain a patent therefor.'[33] Beginning with the decision in *Bilski v Kappos*,[34] the US Supreme Court has issued a series of decisions

---

[31] World Trade Org. Trade-Related Aspects of Intellectual Property Rights https://www.wto.org/english/tratope/tripse/tripse.htm

[32] Nicol et al., above n 20.

[33] 35 USC § 101 (2006).

[34] *Bilski v Kappos* 130 S Ct 3218 (2010) ('*Bilski*').

limiting the scope of patentable subject matter and, in so doing, appears to be going to great lengths to deny patent eligibility for natural product and process patents.

In *Bilski*, the Supreme Court affirmed that some human creativity must be applied to transform natural or mental processes into patentable subject matter. Mere creativity in methods of 'discovering' unpatentable 'laws of nature, physical phenomena and abstract ideas'[35] (this case concerned an abstract idea) is insufficient to endow a discovery falling into these categories with sufficient inventiveness.[36] Confirmation of this approach followed in *Mayo v Prometheus*,[37] where a method for calibrating the correct dosage of a drug to treat a gastrointestinal disorder was held to constitute 'well-understood, routine, conventional activity' that incorporated no transformative step in its conception.[38] The 'machine-or transformation framework'[39] became entrenched as the preferred test for establishing patent eligibility. Having first identified that claims at issue are directed to a patent-ineligible concept, it must *then* be asked whether the patent claims 'transform the nature of the claim' into a patent-eligible application.[40] This constitutes a two-part test that has been adopted in subsequent cases.

The breadth of the exclusion became even more apparent following the Supreme Court's judgment in *Association for Molecular Pathology (AMP) v Myriad Genetics*,[41] where the Supreme Court held the sequencing of naturally occurring DNA (gDNA) to be non-patent-eligible. The insertion of artificial synthesis techniques to create human-made DNA (cDNA) was, however, sufficient to create something 'unquestionably new' and render this form of DNA patent eligible.[42] The basis for this distinction has been criticised as being less than intuitive,[43] and lending little assistance to determining when steps taken during the course of creating product of nature inventions are sufficiently transformative to permit a finding of patent eligibility. The decision did little

---

35 This 'carve out' or exception implicitly contained in § 101 has been recognised repeatedly by the Supreme Court in the cases discussed here.

36 *Bilski*, 3230.

37 *Mayo Collaborative Services v Prometheus Labs, Inc*, 566 US 66 (2012) ('*Mayo*').

38 *Mayo*, 86–7.

39 Joshua Sarnoff, 'Patent Eligible Medical and Biotechnology Patents after *Bilski*, *Prometheus* and *Myriad*' (2011) 19 *Texas Intellectual Property Law Journal* 393, 402.

40 *Mayo* at 86–7.

41 *AMP v Myriad*.

42 *AMP v Myriad*, 2118–19.

43 Dreyfuss et al., above n 29, 560–1.

to allay concerns generated by this line of Supreme Court decisions, increasing uncertainty.[44]

Although not at issue in the Supreme Court proceedings, the Federal Circuit was required to determine the patent eligibility of two forms of method claims: broad methods of diagnosis based on sequence information, and specific methods of diagnoses, based on sequence information.[45] The broad methods comparing a patient's sequence information against 'normal' sequence information were held to be unpatentable on the grounds that they claimed only abstract processes and incorporated no transformative step.[46] The more specific claim, which included the growing of transformed cells *in addition to* comparing growth rates, was patent eligible because it included a transformative step.[47]

And in *Alice Corporation Pty Ltd v CLS Bank*,[48] a computer-implemented scheme for mitigating the settlement risks associated with financial transactions was held to be an abstract idea and thus patent ineligible.[49] Applying the second step of the *Mayo* analysis, the introduction of technological capability to implement the settlement was insufficient to transform the invention and make it markedly different from the fundamental abstract idea.[50]

Importantly, the 'laws of nature' exclusion has been explained on the utilitarian basis that granting a patent would preempt a law of nature, natural phenomena or abstract idea,[51] and impede fundamental innovation.[52] These cases have become the modern scaffold for the preemption framework, in explicitly requiring that a claimed product or process based on a law of nature,

---

44 Robert M Schwartz and Timo Minssen, 'Life after Myriad: the Uncertain Future of Patenting Biomedical Innovation and Personalised Medicine in an International Context' [2015] *Intellectual Property Quarterly* 189.

45 *Association for Molecular Pathology v United States Patent and Trademark Office*, 653 F 3d 1329, 1334 (Fed Cir, 2011); *Association for Molecular Pathology v United States Patent and Trademark Office*, 689 F 3d 1303, 1335 (Fed Cir, 2012).

46 *Association for Molecular Pathology v United States Patent and Trademark Office*, 653 F 3d 1329, 1334 (Fed Cir, 2011), 1355; *Association for Molecular Pathology v United States Patent and Trademark Office*, 689 F 3d 1303, 1335 (Fed Cir, 2012), 1335.

47 *Association for Molecular Pathology v United States Patent and Trademark Office*, 653 F 3d 1329, 1358 (Fed Cir, 2011); *Association for Molecular Pathology v United States Patent and Trademark Office*, 689 F 3d 1303, 1337 (Fed Cir, 2012).

48 *Alice Corporation Pty Ltd v CLS Bank International et al*, 134 S Ct 2347 (2014) (*Alice*).

49 *Alice*, 2357.

50 *Alice*, 2358.

51 See for example *Bilski*, 3231.

52 Lisa Larrimore Oullette, 'Patentable Subject Matter and Nonpatent Innovation Incentives' (2015) 5 *UC Irvine Law Review* 1115, 1121.

natural phenomenon or abstract idea contain a step sufficiently transformative to make it 'markedly different' to that which occurs naturally. And yet, it is not altogether clear that whether or not a particular claim would have a chilling effect on innovation has been (or could be) at the forefront of all of the decisions discussed in this section.[53]

Quite what will be required in a particular scenario has been the subject of some consternation. Subsequent guidance on the application of this test provided by the USPTO[54] did little to alleviate this concern, because it did not define with any precision the meaning of terms such as 'natural products' introduced by the judiciary.[55] Difficulties in application of the judicial tests for patentability, and their focus on whether an invention falls into an exclusion, persisted for a number of years. In short, the Supreme Court decisions on patent eligibility provide little precedential assistance for those seeking to commercialise life sciences research, but conceivably hamstrung lower courts from making findings of patent eligibility in future cases.

3.1.2 Patentable subject matter in Europe

The European approach to assessing patentability for natural products and processes has diverged fundamentally from the US approach, both methodologically and in terms of the outcomes of the cases discussed in the preceding section. The legislative basis for patent eligibility in the EU is contained in the European Patent Convention (EPC).[56] Under this regime, patent ineligibility has developed very differently and has its genesis in the EPC. While Art 52(1) of the EPC provides that patents may be granted in all areas of technology provided the technical requirements for patentability are met, Art 52(2) sets out explicit exclusions and Art 53 provides for exceptions. Art 52(2)(a) provides that 'discoveries, scientific theories and mathematical methods' do not possess the requisite technical requirements to qualify for patentability. This focus on ensuring the invention possesses the requisite 'technical' character means that, unlike the US approach, there is less emphasis on the conceptual basis for the Art 52(2) exceptions.[57]

---

[53] Dreyfuss et al., above n 29, 567–8 discussing *Mayo* and *AMP v Myriad*.

[54] USPTO, Patent Examiners' Manual. The steps to be taken in examining a natural product claim are: Step 1 – ask whether the claim is to a statutory category of claim; Step 2A – ask whether the claim is directed to a law of nature or a natural phenomenon (or an abstract idea where relevant); Step 2B – if Step 2A answered in the affirmative, ask whether the claim recites additional elements that amount to significantly more than the judicial exception [or in other words, transform the claim].

[55] Dan L Burk, 'Dolly and Alice' [2015] *Journal of Law and the Biosciences* 606.

[56] Convention on the Grant of European Patents, 5 Oct. 1973, 1065 UNTS 254.

[57] Siva Thambisetty, 'Alice and "Something More": The Drift towards European Patent Jurisprudence' [2016] *Journal of Law and the Biosciences* 691, 692.

This has resulted in a more liberal attitude to eligibility for certain subject matter. For example, Art 53(b) excepts 'methods for treatment of the human or animal body by surgery or therapy and diagnostic methods practiced on the human or animal body'. This exclusion will apply *only* where four steps have been considered and cumulatively satisfied: (1) an examination phase involving the collection of data; (2) comparison of the results with standard values; (3) identifying any deviance or symptom; and (4) the attribution of this deviance to a particular clinical picture (the deductive phase).[58] Diagnostic methods conducted outside the body and involving an additional technical step will fall outside the exemption and may be patent eligible.[59] The approach presents a considerably more cohesive and less 'atomised' approach than that espoused by the US Supreme Court.[60]

Further, the Directive on the Legal Protection of Biotechnological Inventions (the Biotech Directive),[61] which deals with the patentability of biological materials and processes, explicitly provides for the patentability of isolated DNA sequences if some industrial applicability is disclosed,[62] through disclosure of function and, if relevant, identification of relevant proteins produced.[63] This is justified on the basis that isolation from the human body requires technical processes which nature is incapable of achieving without intervention.[64] This meant that the series of cases[65] involving opposition to Myriad's patents in the EU were in the nature of exercises in statutory interpretation rather than examination of the patent claims to unearth transformative elements.[66] Much of the litigation centred on priority dates for the claims, given that Myriad had amended a number of claims to shore up their claims to patentability. Although the patentability of full DNA sequences was not at issue in the cases (given the

58 Opinion G 1/04 of Dec 16, 2005 (Diagnostic Methods).

59 Opinion G 1/04 of Dec 16, 2005, point 6.2.3.

60 Timo Minssen and Robert M Schwartz, 'Separating Sheep from Goats: a European View on the Patent Eligibility of Biomedical Diagnostic Methods' (2016) *Journal of Law and the Biosciences* 365, 367–8.

61 Directive 98/44/EC of the European Parliament and of the Council of 6 July 1998 on the Legal Protection of Biotechnological Inventions, 1998 OJ (L 213). The Directive has been incorporated into European Law through the EPC *Implementing Regulations to the Convention on the Grant of European Patents* (last amended June 2012).

62 Art 5; pmbl ¶21-24, at L 213/15.

63 Pmbl ¶23-24 at L 213/15.

64 Pmbl ¶21 at L 213/15.

65 Case T 1213/05, *Univ of Utah Research Found v Sozialdemokratische Partei der Schweiz* (EPO Boards of Appeal, 2007); Case T 0666/05 *Univ of Utah Research Found v Institit Curie* (EPO Boards of Appeal, 2008); Case T 0080/05, *Univ of Utah Research Found v Institut Curie* (EPO Boards of Appeal, 2008).

66 Jessica Lai, 'Myriad Genetics and the BRCA Patents' (2015) 5 *UC Irvine Law Review* 1041.

non-contentious nature of this issue under the Biotech Directive), the Board of Appeals confirmed the patentability of a number of inventions, including probes for isolating genes, the detection of mutations and, relevantly, method claims for comparing specific mutations against normal sequence data.[67]

3.1.3 Patentable subject matter in Australia

Australian case law presents us with a different approach again to assessing the patentability of natural products. In order to be patentable, s 18(1) of the Patents Act 1990 (Cth) requires an invention to be a 'manner of manufacture' within the meaning of s 6 of the Statute of Monopolies 1623, or 'a proper subject of letters patent according to the principles which have been developed for the application of s 6 of the Statute of Monopolies'.[68] This threshold test for patentable subject matter will be satisfied where the subject matter in issue is (1) an artificially created state of affairs that (2) has economic utility.[69]

Applied relatively unquestioningly since the test was laid down in 1959, the *NRDC* test was not revisited in any meaningful way until the series of cases involving Myriad's patents came before the courts. At issue in the Australian litigation was the patent eligibility of claims to isolated gDNA and cDNA coding for identified mutations. Methods of diagnosis were not in issue. The High Court of Australia in *D'Arcy v Myriad Genetics Inc (D'Arcy)*[70] unanimously found that the claims in this matter did not constitute patent eligible subject matter. The court applied the *NRDC* test, and held that the two-limb test was not satisfied as there was no 'artificially created state of affairs' as required by the first limb.

The claims were specifically expressed as product claims, but according to the High Court were to informational content only: nothing was 'made', only discerned.[71] There was no outcome brought about by human action, in that the claims did not rely on the changes in chemical composition resulting from isolation.[72] Subsequently, the Australian Patent Office issued a Practice Note[73] which relies on *D'Arcy* in stating that any naturally occurring gene sequence that replicates genetic information is ineligible for patentability; isolation,

---

67 For detailed discussion see ibid, 1057–62.

68 *National Research Development Corporation v Commissioner of Patents* (1959) 102 CLR 252 ('*NRDC*'), 269.

69 *NRDC*, 78–9.

70 *D'Arcy*.

71 *D'Arcy*, [6].

72 *D'Arcy*, [89]–[90].

73 Australian Patent Office, *Examination Practice Following the High Court Decision in D'Arcy v Myriad Genetics Inc*, www.ipaustralia.gov.au/about-us/news-and-community/news/examination-practicefollowing-high-court-decision-darcy-v-myriad.

purification and synthesis are not sufficient to satisfy the manner of manufacture test.[74] This reading of the High Court's conclusion has been questioned on the basis of the unfounded breadth of its interpretation,[75] as have other aspects of the High Court's reasoning that generate uncertainty.[76] In any case, it has been confirmed that the Patent Office does not consider that *D'Arcy* precludes the patentability of 'probes, vectors, methods of production and methods of diagnosis'.[77]

Importantly, the plurality in *D'Arcy* affirmed the use of the *NRDC* test as the correct test to apply in determining patentability.[78] But where claims fall outside the 'established boundaries of patentability', or a 'new class of subject matter', claims must be assessed against a non-exhaustive list of factors, including:[79]

(1) whether granting a patent would have a chilling effect on innovation;
(2) whether granting a patent would impact the coherence of Australian law;
(3) whether a patent would likely be granted in other jurisdictions; and
(4) whether it is more appropriately the place of the legislature to accord patentability.

Although it wasn't necessary to apply these factors to the particular facts of *D'Arcy*, the High Court considered that if they had been required to do so, particularly compelling would have been the fact that allowing the patent would have had a chilling effect on innovation, in that it would present a risk that infringement could take place without the infringer being aware of it.[80]

The High Court was not required to consider method of diagnosis claims in *D'Arcy*, but there is some indication in subsequent lower court cases that method of diagnosis claims remain patentable. In *Meat and Livestock Australia v Cargill*,[81] the patentability of methods for identifying bovine traits from nucleic acid samples was at issue. The claim in this case was to the taking of a sample and analysing that sample to identify components of interest. The

---

74 Charles Lawson, 'Patenting Nucleic Acid Sequences: More Ambiguity from the High Court?' (2018) 25 *Journal of Law and Medicine* 741, 746–7.

75 Mark Summerfield, Proposed Australian Examination Practice Gives Narrow Interpretation to High Court's Myriad's Ruling, Patentology.com.au, 18 Oct 2015, https://blog.patentology.com.au/2015/10/proposed-australian-examination.html

76 See, for example, Lawson, above n 74; Peter McFarlane and Betty Kontoleon, 'Some Legal Issues Regarding the Patenting of Human Genetic Material' (2016) 24 *Journal of Law and Medicine* 181.

77 IP Australia, *Patent Manual of Practice and Procedure* (2017) [2.9.2].

78 *D'Arcy*, [28].

79 *D'Arcy*, [28].

80 *D'Arcy*, [93].

81 *Meat and Livestock Ltd v Cargill Inc* [2018] FCA 51 ('*Cargill*').

High Court had previously held that a new method of using a known drug is patentable, on the basis that pharmaceutical products are patentable and excluding treatments using such products would give rise to an anomaly.[82] However, the broader question of the patent eligibility of physical methods of diagnosis had not been addressed.[83]

Beach J held that this constituted an artificially created state of affairs; it was a claim to methods of using information rather than the practical application of a naturally occurring phenomenon for a particular use.[84] The claims were not at the boundaries of patentable subject matter but fell squarely within the realm of patentability. Relevantly, Beach J refused to entertain applying *Mayo* in the Australian context, stating that '[t]the exposition of the test (particularly the second stage […]) in *Mayo* is too sweeping […] to work out whether [he was] acting consistently or inconsistently with its spirit when determining what it takes to transform an unpatentable law of nature into a patent-eligible application'.[85] Although the decision was appealed,[86] it was not appealed on the question of the correctness of the conclusions on patentable subject matter.

3.2 Impact of Patentable Subject Matter Jurisprudence on the Lo Patent

The new approaches to the determination of patent eligibility outlined above in the US and Australia (and the different approach in Europe) have gone only part of the way in providing an appropriate level of guidance on the boundaries of patentable subject matter, requiring further litigation to bed them down. Litigation relating to the scope of the Lo NIPT patent presented the courts with such an opportunity. However, the decisions that flowed from this litigation actually added little further guidance. Rather, the US, UK and Australian decisions relating to this patent served only to illustrate the breadth of the divergence in interpretation of the subject matter requirement between jurisdictions. The key paragraphs from each of the three key decisions speak for themselves in illustrating just how far apart the jurisdictions were becoming.

In the US, in litigation between Sequenom and Ariosa (among others), the Federal Circuit Court upheld a decision of Illston J in the District Court of California in favour of Ariosa. The Court in *Ariosa Diagnostics, Inc v Sequenom, Inc* (*Ariosa v Sequenom*) found that cffDNA was a naturally occurring phenomenon and that the claimed method of using this phenomenon was

---

82 *Apotex Pty Ltd v Sanofi-Aventis Australia Pty Ltd* [2013] HCA 50 (4 Dec. 2013).
83 Dreyfuss et al., above n 29, 575.
84 *Cargill*, [455].
85 *Cargill*, [492].
86 *Meat and Livestock Australia Limited v Branhaven LLC* [2020] FCAFC 171.

not sufficiently transformative to be patent eligible.[87] Despite the ostensible certainty with which the Court affirmed the lower court decision, however, it has been described by Bob Cook-Deegan and Subha Chandrasekharan as 'grudging'.[88] The following passage from Reyna J's judgment crisply encapsulates the reason why the court really had no option but to hold the patent ineligible, based on Supreme Court jurisprudence:

> It is undisputed that the existence of cffDNA in maternal blood is a natural phenomenon. Sequenom does not contend that Drs. Lo and Wainscoat created or altered any of the genetic information encoded in the cffDNA, and it is undisputed that the location of the nucleic acids existed in nature before Drs. Lo and Wainscoat found them. The method ends with paternally inherited cffDNA, which is also a natural phenomenon. The method therefore begins and ends with a natural phenomenon. Thus, the claims are directed to matter that is naturally occurring.[89]

The problem for Sequenom in defending the patent was that the claims in issue faced the same problems as in *Mayo* and *Myriad:* they were 'preemptive broad claims'.[90] As rightly pointed out by Cook-Deegan and Chandrasekharan, the much needed clarification on the nuances of patent eligibility would have to wait until the courts were given the opportunity to consider these questions for claims that did not suffer such faults.[91] The Supreme Court was certainly not about to enter the fray once more, refusing to grant certiorari to hear an appeal from the Federal Circuit decision.[92]

In contrast with these difficulties with the claims as drafted in the US, the UK High Court encountered no such problems, holding that similar claims were mostly valid and to some extent infringed.[93] In this case a number of patents were under challenge, including the Lo patent. In exploring the eligibility of subject matter in the Lo patent claims, Carr J focused on the provision

---

87 *Ariosa Diagnostics, Inc. v Sequenom, Inc*, 788 F 3d 1371 (Fed Cir, 2015) ('*Ariosa v Sequenom*').

88 Robert Cook-Deegan and Subhashini Chandrasekharan, 'Sequenom v. Ariosa – The Death of a Genetic Testing Patent' (2017) 375 *New England Journal of Medicine* 2418, 2418.

89 *Ariosa v Sequenom*, pages 8–9 of Reyna J's slip opinion.

90 Cook-Deegan and Chandrasekharan, above n 88, 2419.

91 Ibid.

92 Patent Docs, *Supreme Court Denies Certiorari in Sequenom v. Ariosa* (26 June 2016), www.patentdocs.org/2016/06/supreme-court-denies-certiorari-in-sequenom-v-ariosa.html.

93 *Illumina, Inc v Premaitha Health Plc* [2017] EWHC 2930.

in the UK *Patents Act* excluding claims relating to discoveries as such. His Honour concluded:

> I do not accept that, properly construed, claim 1 is a claim to a discovery as such. The claims are not directed to information about the natural world, but rather to a practical process, namely a 'detection method' which uses information about the natural world. Claim 1 is directed to the detection of foetal DNA in a sample of plasma or serum. Such samples do not exist in the natural world and must be artificially created. The claimed method of detection is also an artificial process which does not exist in the natural world. The claim is to a practical process of implementing a discovery, for practical applications. The actual contribution, as a matter of substance, does not fall solely within the excluded subject matter and is technical in nature.[94]

In Australia, in June 2016, Sequenom started infringement proceedings in the Australian Federal Court relating to the Lo patent and other NIPT patents. The action was brought against Sonic Healthcare, Australian Clinical Labs and Ariosa in relation to their prenatal NIPT assay, the Harmony test. Sequenom claimed that the supply of the Harmony test by Ariosa, Sonic Healthcare and Australian Clinical Labs infringed its patent. In August 2016 the respondents filed cross-claims for invalidity of Sequenom's patent. Included in the grounds was lack of patentable subject matter.[95]

The Australian litigation gave the Federal Court the opportunity to consider for itself how to apply the manner of manufacture test to method claims grounded in naturally occurring phenomena. Beach J at first instance held the patent to be mostly valid and infringed (save for one claim).[96] The rationale for his decision was that:

> The Patent does not simply claim the discovery of cffDNA in maternal blood. Rather, it claims a new and inventive practical application of the discovery comprising a method requiring human action to detect, in an artificially created sample of maternal plasma or serum, a DNA sequence as being of fetal rather than maternal origin. And prior to the invention, no-one had worked or was working a method comprising the detection of cffDNA in plasma or serum samples extracted from pregnant females.[97]

Beach J's first instance decision was upheld on appeal, with the court similarly holding that 'Unlike the position in Myriad, claim 1 is not, as a matter of sub-

---

94 *Illumina, Inc v Premaitha Health Plc* [2017] EWHC 2930.

95 DLA Piper LLP, 'The Patentability of Diagnostic Methods in Australia the Australian Sequel to Sequenom, Myriad and Mayo', Lexology (24 Oct. 2016) www.lexology.com/library/detail.aspx?g=4230dae4-843e-4925-a015-fe64c8c68804.

96 *Sequenom, Inc. v Ariosa Diagnostics, Inc.* [2019] FCA 1011.

97 *Sequenom, Inc. v Ariosa Diagnostics, Inc.* [2019] FCA 1011, [595].

stance, directed to genetic information, but to a method involving the practical application of a means for identifying and discriminating between maternal and foetal nucleic acid'.[98]

In summary, then, to date what we have is three very different decisions on essentially the same patent claims. As the Full Court of the Federal Court aptly put in in *Ariosa v Sequenom – Australia*:

> It is appropriate to note that there is no real prospect of falling out of step with the international community of patent decisions in reaching this conclusion, because there are disparate approaches taken abroad to the question of patentable subject matter, involving different legal tests, none of which is in the same form as that which has developed in Australia.[99]

Although the UK and Australian courts get to the same end point, their line of reasoning is somewhat different. In contrast, the outcomes in the US and Australia, which have similar statutory requirements for subject matter eligibility, are remarkably divergent.

4. MORE RECENT DEVELOPMENTS IN THE NIPT PATENT LANDSCAPE

The question, then, is whether these different decisions make any difference, either to global or to local delivery of NIPT. The Lo patent family has now expired, the technology has advanced, and there are many more patents. As such, the fact that the Lo patent was invalidated in the US and some claims were lost elsewhere hardly matters to the development and delivery of NIPT from this perspective at this moment in time. However, if the legacy of *Mayo*, *Myriad* and other cases in the US is that the bounds of patentable subject matter continue to contract, this will inevitably have consequences for this rapidly developing healthcare service. Moreover, if the divergence in patent law continues to manifest itself, then we may see a shift in business investment from the US to more patent-friendly regions of the world. And if the competition between the multiple players in the industry continues, we may see much expenditure on litigation with little gain in terms of enhanced NIPT service delivery and patient access.

Perhaps it is fortunate that the US courts are finding ways to circumvent the perceived harshness of the Supreme Court decisions, in the sense that these decisions have resulted in many patents being struck down by the courts.

---

[98] *Ariosa Diagnostics, Inc v Sequenom, Inc* [2021] FCAFC 101, [155] ('*Ariosa v Sequenom – Australia*').

[99] *Ariosa v Sequenom – Australia*, [161].

Patent prosecution analysis by Mateo Aboy and his colleagues has revealed that it is not just patents claiming subject matter closely aligned to the subject matter that are falling foul of the Supreme Court in *Mayo* and *AMP v Myriad*, but a wide range of other patents in the fields of biotechnology, diagnostics and personalised medicine as well.[100]

The first in this line of notable decisions that illustrate the possible options for distinguishing the Supreme Court decision in *Mayo* was *Vanda Pharmaceuticals, Inc v West-Ward Pharmaceuticals International Ltd* (*Vanda*).[101] This case concerned a method for treating schizophrenia based on the genotype of a patient: patients with a particular genotype associated with a poor drug metabolism are known to be at increased risk of cardiac arrest if higher dosages of the drug iloperidone are administered. The Court of Appeals for the Federal Circuit found the claims patent eligible on the basis that unlike the claims in *Mayo*, the claims were not directed toward a natural relationship (Step 1A), but rather to a novel application of the natural relationship to the treatment of a disease.[102] In other words, there was no invocation of Step 2B, which is required to be considered where subject matter is classified as constituting laws or products of nature. *Vanda* thus seems to have tempered the impact of *Mayo* to some degree.

This impact of the decision was clarified in a USPTO examination guidance memo (7 June 2018),[103] which confirmed that method of treatment claims pass the Step 2A inquiry, and that subsequent consideration of Step 2B is not required for such claims.[104] Subsequent revisions to the USPTO's guidance on patent eligibility further cemented this position, which may broaden the chasm between methods of medical treatment applying natural relationships (which are patent eligible) and methods of diagnosis (which are not).[105]

In *Rapid Litigation Management Ltd v CellzDirect, Inc* (*CellzDirect*),[106] at issue was a new technique for cryogenically freezing hepatocytes, or liver

---

[100] Aboy et al., above n 29; Mateo Aboy et al., '*Mayo*'s Impact on Patent Applications related to Biotechnology, Diagnostics and Personalized Medicine' (2019) 37 *Nature Biotechnology* 513.

[101] *Vanda Pharm Inc v West-Ward Pharm Int'l Ltd*, 887 F.3d 1117 (Fed Cir, 2018) ('*Vanda*').

[102] *Vanda*, page 29 of Lourie J's slip opinion.

[103] Robert W Bahr, Recent Subject Matter Eligibility Decision: *Vanda Pharmaceuticals Inc v West-Ward Pharmaceuticals* (USPTO, Memorandum to the Patent Examining Corps; 7 June 2018) www.uspto.gov/sites/default/files/documents/memo-vanda-20180607.PDF.

[104] Ibid.

[105] www.uspto.gov/sites/default/files/documents/pegoct2019update.pdf.

[106] *Rapid Litigation Management Ltd v CellzDirect, Inc*, 827 F.3d 1042 (Fed Cir, 2016).

cells, for later use in testing, diagnosis and treatment. It improved on earlier techniques in that the viability of thawed cells was vastly improved. Again, the Court of Appeals for the Federal Circuit found it unnecessary to progress to *Mayo*'s Step 2B, finding that the claims were 'directed to a new and useful laboratory technique for preserving hepatocytes'.[107] The claim did not simply constitute the discovery of a law or product or nature. Rather, the inventors employed their discovery that hepactocytes could survive multiple freeze cycles, to create a new and improved method of preservation.

In yet another case where the Court of Appeals for the Federal Circuit was required to consider the seemingly narrow test in *Mayo*, *Endo Pharm Inc v Teva Pharm USA, Inc* (*Endo*),[108] the Court took an almost identical line to that taken in *Vanda*. The claims in question related to a method of administering a reduced dose of oxymorphone, a pain relief drug, to patients with impaired kidney function. Again, the Court determined that this was a claim that 'focuse[d] solely on step one' (referred to as Step 2A in the *Mayo* analysis above); that the claims were not directed to patent-ineligible subject matter in the sense of a law or product of nature, but were instead directed toward a patent-eligible method of using oxymorphone to treat pain in a renally impaired patient.[109] Like *Vanda*, this method relied on a dosage regimen based on the result of conducting tests on patients; whereas the method in *Vanda* relied on genetic testing, the method in *Endo* relied on kidney function testing.

This snapshot of cases confirms that claims framed as method of treatment claims (as opposed to method of diagnosis claims) are far more likely to evade consideration under Step 2B of the Mayo analysis. Very few claims that have required consideration under Step 2B have been upheld.[110] What the cases do seem to indicate is a willingness on the part of the Court of Appeals for the Federal Circuit to apply the *Mayo* test in a manner that favours the interpretation that a particular method is a method of treatment or method of preparation. What it may also show is that claims are being drafted and interpreted in a way that favours these types of interpretation over an interpretation that they represent a method of diagnosis,[111] which is far more likely to fall foul of the *Mayo* two-step test.

---

[107] *CellzDirect*, page 8 of Prost CJ's slip opinion.

[108] *Endo Pharmaceuticals Inc v Teva Pharmaceuticals USA, Inc*, 919 F.3d 1347 (Fed Cir, 2019) (*Endo*).

[109] *Endo*, pages 9–11 of Stoll J's slip opinion.

[110] Dreyfuss et al., above n 29, 562–3.

[111] Mateo Aboy et al., 'One Year after Vanda, Are Diagnostics Patents Transforming into Methods of Treatment to Overcome Mayo-Based Rejections?' (2020) 38 *Nature Biotechnology* 279.

One effect of *Mayo*, however, is that the decision has provided District Courts with considerable judicial discretion to invalidate patent claims based on a subjective evaluation of their eligibility, given the uncertain boundaries of the narrow test for patent eligibility laid down by the Supreme Court.[112] Of note is the fact that the Federal Court in *CellzDirect* and *Endo* reversed the respective District Court findings of invalidity, which demonstrates the conservative application by District Court judges of the *Mayo* test. This is also playing out at Patent Office level, where a far greater proportion of patent applications are being rejected as being patent ineligible in the US than in other jurisdictions.[113]

One District Court judge who has exhibited significant discretion in *invalidating* claims in the NIPT space is Justice Susan Illston, the District Court judge who heard *Ariosa v Sequenom*. As will be recalled, Illston J invalidated Sequenom's patents and this decision was upheld by the Court of Appeals for the Federal Circuit. In a subsequent decision, *Illumina, Inc v Ariosa Diagnostics, Inc*,[114] Illston J was required to consider the validity of two patents in a suite of patents that Illumina and Sequenom had previously agreed to pool.[115] The '751 patent comprised procedures to separate foetal and maternal DNA in a maternal blood sample, while the '931 patent involved procedures to separate foetal DNA from a maternal sample through 'size discrimination methods'. Not surprisingly, Illston J held the claims to be patent ineligible, based on the fact they were representative of a natural phenomenon (naturally occurring foetal DNA) that were not transformed into something new by virtue of the patent claims.

Her Honour held that there was nothing in the patents as claimed 'to overcome the "directed to" inquiry'.[116] In her comparison with other case law, her Honour found that the facts at hand were more analogous to *Ariosa v Sequenom* than to *CellzDirect* for the following reasons:

> In CellzDirect, the end result was cryogenically frozen useful liver cells that did not occur in nature. In Ariosa, as is the case here, the claims are directed to a testable quantity of genetic information found in nature. Unlike CellzDirect, the end result is

---

[112] Christopher M Holman, 'Patent Eligibility Post-Myriad: A Reinvigorated Judicial Wildcard of Uncertain Effect' (2014) 82 *Wash L Review* 1796, 1798.

[113] Kevin Madigan and Adam Mossof, 'Turning Gold into Lead: How Patent Eligibility Doctrine is Undermining US Leadership in Innovation' (2017) 24 *George Mason Law Review* 939.

[114] *Illumina, Inc v Ariosa Diagnostics, Inc* (N.D. Cal. 2018).

[115] Hawkins et al., above n 11, 442.

[116] *Illumina, Inc v Ariosa Diagnostics, Inc* (N.D. Cal. 2018), page 10 of Illston J's slip opinion.

> naturally occurring. Accordingly, the Court finds plaintiffs' arguments unpersuasive and holds that the patents are directed to patent ineligible concepts.[117]

In contrast to the situation in *Ariosa v Sequenom* some five years earlier, where the Federal Circuit upheld Illston J's decision, in this case the appeal court rejected her Honour's reasoning, preferring to align their decision more with *CellzDirect* than with *Ariosa v Sequenom.*[118] Lourie J gave the opinion of the Court in *Illumina, Inc v Ariosa Diagnostics, Inc* (*Illumina v Ariosa*), with a dissenting opinion by Reyna J. Importantly, Lourie J characterized the case as being not about diagnostics (as in *Mayo* and *Ariosa v Sequenom*) or methods of treatment (as in *Vanda),* but about a method of preparation (as in *CellzDirect*).[119] His Honour agreed with Illumina's characterisation of the natural phenomenon in issue: 'that cell-free fetal DNA tends to be shorter than cell-free maternal DNA in a mother's bloodstream.'[120] This led the court to conclude that 'the claims are *not* directed to that natural phenomenon but rather to a patent-eligible method that utilizes it'.[121]

The crucial point of distinction from *Mayo*, *Ariosa v Sequenom* and like cases was that the claims in issue in those cases were squarely directed to detecting a natural phenomenon. Here, in contrast, the court characterised the claims as being directed towards *exploiting* a natural phenomenon, namely the natural size differentials between maternal and foetal cffDNA so that the foetal DNA can be analysed for chromosomal aberrations.[122] Although not directly aligned with *CellzDirect*, Lourie J concluded that it was sufficiently instructive. In essence, his reasoning was that the claims in both cases were not simply directed towards observing or detecting natural phenomena but sought new and useful methods of utilising natural phenomena.[123]

In contrast, Reyna J, in dissent, was of the view that the claims were directed towards the 'surprising discovery of size discrepancy of cff-DNA in a mother's blood – an undisputed natural phenomenon'.[124] In his Honour's view, claim 1 of the '751 patent starts with extracting a sample of blood plasma or serum from a pregnant mother, then fractionating it and analysing the components of

---

[117] *Illumina, Inc v Ariosa Diagnostics, Inc* (N.D. Cal. 2018), page 11 of Illston J's slip opinion.

[118] *Illumina, Inc v Ariosa Diagnostics, Inc* 952 F.3d 1367 (Fed Cir, 2020), modified and reissued following a petition for rehearing filed by Defendants-Appellees, 967 F.3d 1319 (Fed Cir, 2020).

[119] *Illumina v Ariosa* (reissued opinion), page 8 of Lourie J's slip opinion.

[120] *Illumina v Ariosa* (reissued opinion), page 10 of Lourie J's slip opinion.

[121] Ibid.

[122] *Illumina v Ariosa* (reissued opinion), page 14 of Lourie J's slip opinion.

[123] Ibid.

[124] *Illumina v Ariosa* (reissued opinion), page 3 of Reyna J's slip opinion.

the fraction.[125] In this sense, then, the methods involved a natural phenomenon and were directed to a natural phenomenon. According to Reyna J, there was no 'claimed advance' over the natural phenomenon.[126]

As noted by Kevin Noonan in his discussion of the first of the two Federal Circuit opinions in this case in his Patent Docs blog, this illustrates that there is not yet unanimity in the judiciary on the circumstances in which the use of a natural phenomenon or abstract idea can cross the subject matter threshold.[127] Indeed, this is far from the case. It is interesting to note, at this point, that Reyna J, who dissented in *Illumina v Ariosa*, gave the opinion of the court in *Ariosa v Sequenom*. This indicates that he is at least one member of the Federal Circuit who is likely to remain unpersuaded that claims embracing ineligible subject matter could ever cross the eligibility threshold. As Noonan points out, Reyna J and other judges sitting on both lower and higher courts in the US, follow 'the conventional view that applying routine, well-understood, and conventional methods to a newly discovered natural phenomenon is enough, *per se*, to preclude patent eligibility'.[128]

As such, it would perhaps be foolhardy to suggest that the US judiciary as a whole has embraced the creative approaches being used by the Federal Circuit to distinguish methods of treatment and preparation from methods of diagnosis. Despite this reluctance on the part of some members of the judiciary, however, we are seeing a consistent trend in recent decisions of the Federal Circuit to move toward acceptance of such claims. Given the apparent reticence of the Supreme Court to accept more patent subject matter cases,[129] some resilience does appear to be emerging within the US patent system. We posit that this is important, if the patent system is to withstand calls by industry for Congress to reduce the subject matter requirement to a mere formality.[130]

In contrast with the experience in the US, in the UK the NIPT patent disputes remain firmly embedded in more traditional patent doctrine. This is illustrated in the latest UK NIPT case, relating to a patent claiming non-invasive

---

125 *Illumina v Ariosa* (reissued opinion), page 5 of Reyna J's slip opinion.

126 *Illumina v Ariosa* (reissued opinion), page 6 of Reyna J's slip opinion.

127 Kevin E Noonan, *Illumina, Inc. v Ariosa Diagnostics, Inc.* (Fed. Cir. 2020) (17 March 2020) www.patentdocs.org/2020/03/illumina-inc-v-ariosa-diagnostics-inc-fed-cir-2020.html?utm_source=feedburner&utm_medium=email&utm_campaign=Feed%3A+PatentDocs+%28Patent+Docs%29.

128 Ibid.

129 In January 2020, for example, the Supreme Court declined to issue a writ of certiorari in *Vanda:* www.supremecourt.gov/search.aspx?filename=/docket/docketfiles/html/public/18-817.html.

130 Dreyfuss et al., above n 29, 4, 31; Congressional Research Service, Patent-Eligible Subject Matter Reform in the 116th Congress (2019), 27, https://fas.org/sgp/crs/misc/R45918.pdf.

detection of foetal genetic traits (aligned with the US '931 patent which featured in *Illumina v Ariosa*).[131] This patent was held by Arnold J to be valid and infringed. Much of the judgment focused on the question of whether there was an inventive step and insufficiency, familiar territory for most patent litigators. It was only at the end of the judgment that attention turned to patentable subject matter. The difficulty for the defendant was that the patent was not open to challenge on this ground because of the earlier *Illumina* case decided by Arnold J and other decisions. What the defendants were seeking to do was to open the door to challenge in a higher court, by asking the court to make findings of fact. Arnold J rejected the invitation to do so, holding:

> The Defendants contend that, even if otherwise valid, the claimed inventions are excluded from patentability under section 1(2)(a) of the 1977 Act on the ground that they are discoveries. The Defendants accept, however, that, as the law presently stands, this contention cannot succeed. They ask the Court to make appropriate findings of fact to enable the issue of law to be argued in a higher court if necessary. [...] I agree with the Claimants that it is unsatisfactory for the Court to be asked to make findings of fact in a legal vacuum. [...] Accordingly, I decline to make the findings requested.[132]

Although both the obviousness and insufficiency grounds for invalidity brought by the defendants were ultimately unsuccessful, what this judgment shows is that there is ample scope for raising such arguments in NIPT patent disputes.

5. WHERE DOES THIS GET US?

This analysis shows that in the US the flurry of judicial decisions striking down patents based on their failure to satisfy of the patentable subject matter requirement appears to have abated, at least to some extent, and at least for biotechnology-related subject matter.[133] The courts have found ways to circumvent some of the apparent harshness resulting from the rigid application of the *Bilski–Mayo–AMP–Alice* series of Supreme Court cases. It is perhaps overly optimistic, however, to expect that there will be universal acceptance within the US judiciary of the reasoning behind the decisions in *Vanda*, *CellzDirect* and *Illumina v Ariosa*. The opinions of Illston J at first instance, and Reyna J in dissent in the Federal Circuit in *Illumina v Ariosa*, illustrate this point. Yet

---

[131] *Illumina, Inc v TDL Genetics Ltd* [2019] EWHC 1497.

[132] *Illumina, Inc v TDL Genetics Ltd* [2019] EWHC 1497, [242] and [245].

[133] But see *American Axle & Manufacturing v Neapco Holdings LLC,* 927 F.3d 1333 (Fed. Cir. 2019) for a recent example of the Federal Circuit's approach regarding other subject matter.

there appears to be no sense of urgency on the part of higher authorities to short-circuit further jurisprudential development of this line of authority. The Supreme Court appears reluctant to consider new cases, even in light of these new Federal Circuit decisions. Congress, moreover, seems to be in no hurry to amend the legislative provisions relating to subject matter.[134]

The primary outcome from this analysis of key NIPT patent cases is that it illustrates that there is resilience within the patent system (via its interpreters, the judiciary). Patent law will inevitably operate in a state of flux whenever the judiciary is forced to come to terms with seismic technological shifts and grapple with fundamental issues of patentability. As we have seen, the most recent challenge for the courts has been to deal with the elision between the patent ineligibility of natural phenomena and abstract ideas, on the one hand, and the need for patent protection within the biotechnology and information technology industries on the other. The approach of the Supreme Court in dealing with this challenge is widely accepted to have been heavy handed and lacking in nuance and detail. It was the catalyst for a period of uncertainty for all stakeholders.

Knee-jerk reactions from particular sectors, demanding law reform in response to what they perceive as adverse judicial decisions, are not the appropriate way to deal with these issues. Patent law in the US has been through a period of upheaval characterised by significant legislative and judicial activity. If proposals for additional reform to address the Supreme Court's innovation-unfriendly approach were accepted by Congress, it is highly likely that this would lead to even greater periods of uncertainty while the administration and courts interpret new statutory provisions. The other risk is that such changes could have the effect of tilting the scales too far in the other direction. The result could be a more liberal patent environment than that which existed prior to the Supreme Court interventions, beyond simply unpicking the work of the judiciary. Irrespective of whether or not it was the 'correct' decision to hold the claims in *Illumina v Ariosa* to be patent eligible, the decision may have the positive effect of dampening calls for further transformational patent reform.

One possible consequence of these recent developments in the US is that the eligibility of patents relating to NIPT and other biotechnology-related subject matter may become less divergent globally. However, we do not suggest that the patentable subject matter requirement is now completely uniform or harmonised across jurisdictions. The subject matter inquiry in Europe remains very different from that in the US. The recent Australian Full Court decision in

[134] See story 2 in Donald J Zuhn's 'Top Four Stories for 2020' (2021), www.patentdocs.org/patent_legislation/.

Ariosa v Sequenom – Australia indicates that, while the language of the subject matter inquiry is similar to that in the US, the outcomes are quite different. The Australian courts remain resolute in their determination to develop their own line of jurisprudence. If anything, the outcomes are more closely aligned with the outcomes of UK jurisprudence, considering very different subject matter requirements, than with the early US decisions in *Mayo* and *Ariosa v Sequenom*. Admittedly, the recent US decisions may encourage litigants and the courts to return to the more familiar territory of novelty, inventive step and scope of claims in determining patent validity. The ways in which the traditional criteria relating to novelty, inventive step and scope of claims are applied in NIPT patent cases will still, of course, vary, depending on national nuances. Nor would we say that the patent system won't be exposed to similar perturbations in the future.

Recognising that the patent system is imperfect, is there anything that can be done during periods of technology upheaval to ensure the engines of innovation keep moving? In both the biotechnology and the pharmaceutical industries, patents play a critical role, given the significant costs of product development. It might be argued that different considerations apply to the field of diagnostics where capital development costs are lower. But this does not reflect the reality for NIPT, given that it took some ten years from the research of Dennis Lo and others to translate into market-ready products. In this regard, then, perhaps NIPT is distinct from other forms of diagnostic testing. It certainly is the case that NIPT is very much a commercial venture and thus a lucrative technology, rather than being primarily the domain of public sector testing facilities and hospitals. This might help to explain the protracted litigation that characterises the NIPT landscape.

A number of mechanisms outside of patent eligibility exist to deal with challenges relating to progress and technological development. However, the issue of patent eligibility is such a fundamental one that cohesion across jurisdictions is of critical importance. Over the past decade, jurisdictional divergence on the patent eligibility question has intensified. But the NIPT litigation may indicate an attempt to temper the narrow line taken by the US Supreme Court (and its lower relatives, the District Courts and Courts of Appeal of the Federal Circuit) over the course of the past decade. It supports the conclusion that although we are not yet at a point where cohesion between jurisdictional interpretations of patent eligibility is evident, there is at least some consistency.

There is evidence in the NIPT field that those who would seek to provide this service are more concerned about the risk of patent infringement than are those who provide diagnostic genetic services.[135] Given the ubiquity of pregnancy and the importance of promoting the health and welfare of the next generation, access to NIPT is perhaps even more important from the equity perspective than other areas of diagnostic testing. This is illustrative of the important need to fix the boundaries of patent eligibility for the benefit of society as a whole.

There is some urgency in addressing these issues. We have seen that patent law has the capacity to evolve and adapt to technological change. It is sufficiently resilient to provide responses to the challenges presented by new technologies. But this is often a lengthy exercise. In the 1980s the first gene patents were granted: it has taken many years to resolve fundamental questions about patent availability in this field. The wheels of patent law do move inexorably more slowly than the wheels of technology.[136] This illustrates the critical need for adequate flexibility within the concept of patent eligibility to permit it to respond efficiently to rapid technological innovation. The NIPT litigation provides a useful case study that illustrates judicial willingness and capacity to issue such a response.

---

[135] Dianne Nicol, Jane Nielsen and Verity Dawkins, *D'Arcy v Myriad Genetics: The Impact of the High Court's Decision on the Cost of Genetic Testing in Australia* (Hobart: Centre for Law and Genetics Occasional Paper No. 9, 2018), 65; Naomi Hawkins, 'Patents and Non-invasive Prenatal Testing: Is there a Cause for Concern' [2020] *Science and Public Policy* 1.

[136] Dianne Nicol, 'The Pendulum of Patents, Principles and Products – from the Industrial Revolution to the Genetic Revolution', in press, *Kritika*.

4. Patentability of biofabricated human organs: 'products of nature' or 'products derived from nature' revisited

Pheh Hoon Lim and Phoebe Li[1]

1. INTRODUCTION

In recent years, progress in 'biofabrication' has offered new ways to realise the promise of mass production of human organs, with the potential of relieving the burden of organ shortage. Biofabrication is a broad concept encompassing various types of methods used to manufacture tissue engineering and regenerative medicine. It can be understood as comprising two main process methods: 3D bioprinting and bioassembly.[2] In this chapter we take a 3D biofabricated human organ as an example in assessing the patent eligibility of 'products' of biofabrication.

The development of a 3D biofabricated organ starts by growing human cells from biopsies or stem cells[3] as bioink fed into the printers to create three-dimensional replicas. Currently there are limited clinical applications due to limitations on the vascularity of biofabricated organs.[4] Successful

---

1 Phoebe Li is funded by the Leverhulme Trust 'Biomodifying technologies: governing converging research in the life sciences' [RPG-2017-330], and the UK ESRC 'Biomodifying technologies and experimental space' [ES/P002943/1].

2 Jürgen Groll et al. 'Biofabrication: reappraising the definition of an evolving field' (2016) *Biofabrication* 8, 013001. 3D bioprinting refers to the direct spatial arrangement of cells and materials; bioassembly has a similar effect but uses automated assembly of cell containing building blocks.

3 James O'Toole '3-D-printed organs are on the way' *CNNMoney* (New York) 4 November 2014, https://money.cnn.com/2014/11/04/technology/innovationnation/3d-printed-organs/ accessed 3 December 2020.

4 Suradip Das et al. 'Innervation: the missing link for biofabricated tissues and organs' (2020) Regenerative Medicines 5(11), www.nature.com/articles/s41536-020-0096-1 accessed 27 August 2021.

clinical applications are primarily around thin layers of biofabrication, such as skin used for wound reconstruction, as opposed to complicated organs for transplantation, such as a liver or kidney – future breakthroughs in the vascularity of biofabricated organs would be required for the latter.[5] While offering hopes of customised patient treatment such as organ replacement, cell regeneration and access to affordable healthcare, creation of a 3D biofabricated organ from human samples of tissues however challenges the patent landscape. As the biological components of such biofabricated organ products are already in existence in nature,[6] concern is raised over the blurring of the distinction between discoveries and inventions.

Alongside the hopes of possible invention of human organs that can overcome tissue rejection in the human body is a sordid reminder that the sale of body parts such as the liver or kidney takes place. Organised crime in this area preys on helpless victims or those vulnerable to the barter of such organs for cash, while providing at the same time access to others able to afford these organs' purchase. The 1984 initiative in China to legitimise organ removal from death row prisoners to overcome a shortage of human organs for transplant purposes drew public consternation,[7] as it lent itself instead to a lucrative trade reaping a 'staggering US$1 billion' per annum amid reports that China's state-run hospitals were involved in the trafficking of human organs.[8]

Putting aside the black market trade of human organ trafficking and the blatant commercialisation and disregard of ethical considerations, the need for organs available for transplant is a pressing problem in the healthcare industry due to waiting list numbers.[9] In the broad context of organ trafficking and access to health technologies, this chapter examines the implications for

---

[5] Stephanie Ishak and Shari Lipner 'A review of 3-dimensional skin bioprinting techniques: applications, approaches, and trends' (2020) 46 (12) Dermatologic Surgery 1500-1505.

[6] Craig C Martin and Sara Tonnies Horton 'Patent eligibility of 3D printed organs will soon be an issue' *Today's General Counsel* June/July 2014, https://jenner.com/system/assets/publications/13151/original/martin_horton_Todays_GC_2014.pdf?1404161316 accessed 3 December 2020.

[7] JD Briggs 'The use of organs from executed prisoners in China' (1996) 11(2) Nephrol Dial Transplant 238–40.

[8] John Carney 'A human harvest: China's organ trafficking exposed in shocking documentary that alleges the illegal trade is now worth a staggering US$1 billion a year' *Daily Mail* (Australia, 7 April 2015), www.dailymail.co.uk/news/article-3027088/A-Human-Harvest-Chinas-organ-trafficking-exposed-shocking-documentary-alleges-illegal-trade-worth-staggering-1-billion-year.html accessed 3 December 2020.

[9] Data from the United States Department of Health and Human Services shows 91,800 and 12,094 patients awaiting kidney and liver transplants respectively as at 3 December 2020, https://optn.transplant.hrsa.gov/data/. According to the United States National Kidney Foundation the average waiting time is more than 3.6 years, www

biofabricated human organs arising from judicial dissonance over the patentability of genes and genetic material in different legislative regimes.

Based on the two main strands of reasoning presented by (1) the scenario on bioproducts in the *Oncomouse*[10] and *Dolly the Sheep*[11] cases, and (2) the debate on 'isolated genes' in the *Myriad* saga,[12] we query whether a 3D biofabricated human organ (a product created by the bioprinting and multilayering of naturally occurring cells) is a man-made invention. Or, arguably, is it a non-patentable product of nature despite the need for human intervention to create it? Drawing a parallel between isolated DNA (and bioproducts of higher life forms/mammals) and a biofabricated human organ, we contemplate further whether the latter could satisfy the patent eligibility threshold when the essential elements of the claimed biological compositions do not alter the fundamental quality of an organ as it exists in the body.

In this chapter we focus our investigations on the patentability of biofabricated 'products' (organs). There are other issues around the patentability of the 'processes', such as the production methods and the transplantation of biofabricated organs, which relate to the medical exception in patent law. As that is a substantial topic in its own right, previously published on by one of these authors,[13] it will not be covered in the discussion here.

This chapter comprises six sections. In section 2 we review what is *not* an invention, while in section 3 we focus on the patentability issues of bioproducts, exemplified by the *Oncomouse* and *Dolly* cases. We then examine the *Myriad* saga on the patentability of isolated DNA and genes in section 4, for a comparative analysis. In section 5 we assess the patent eligibility of biofabricated organ products. This is followed by our concluding remarks, in the last section.

.kidney.org/news/newsroom/factsheets/Organ-Donation-and-Transplantation-Stats accessed 3 December 2020.

[10] United States Patent and Trademark Office (USPTO) Patent 4,736,866 (12 April 1988); European Patent 0 169762 (filed in 1985) and *Harvard College v Canada (Commissioner of Patents)* [2002] 4 SCR 45 (Supreme Court of Canada 5 December 2002).

[11] *In re Roslin Inst. (Edinburgh)* 750 F 3d 1333 (Fed Cir. 8 May 2014).

[12] See United States: *Association for Molecular Pathology v USPTO* 702 F Supp 2d 181 (SDNY 29 March 2010); *Association for Molecular Pathology v USPTO* 689 F 3d 1303 (Fed Cir. 16 August 2012) and *Association for Molecular Pathology v Myriad Genetics Inc* 133 S Ct 2107 (13 June 2013). See Australia: *Cancer Voices Australia v Myriad Genetics Inc* [2013] FCA 65 (15 February 2013); *D'Arcy v Myriad Genetics Inc* [2014] FCAFC 115 (7, 8 August 2013, 5 September 2014) and *D'Arcy v Myriad Genetics Inc* [2015] HCA 35 (7 October 2015).

[13] Phoebe H Li '3D Bioprinting technologies: patents, innovation and access' (2014) 6(2) LIT 282. DOI:http://dx.doi.org/10.5235/17579961.6.2.282.

2. PATENT LANDSCAPE: WHAT IS NOT AN INVENTION?

Biotechnical inventions have challenged the phenomena of patenting life forms, based on the classical view that patent law does not, to a great extent, concern itself with living material. In the discovery and invention debate, courts in the United States spell out that laws of nature, physical phenomena and abstract ideas (which include products of nature in the public domain) are not patent eligible subject matter.[14] However, variations of such naturally occurring organisms to create new organisms may be patent eligible. Legislation in the United Kingdom and European Union states very specifically that discoveries, scientific theories and/or mathematical methods are not inventions.[15]

Patent law in Australia (and New Zealand) defines an invention as a 'manner of manufacture' within the meaning of section 6 of the Statute of Monopolies.[16] The celebrated 1959 judgment of the High Court of Australia in *National Research Development Corporation v Commissioner of Patents (NRDC)*[17] remains a definitive statement on the question of patentable subject matter to this day.[18] The patent claimed a method on the use of a known herbicide for eradicating weeds effectively without harming the good crop, compared with earlier compositions that wiped out both weeds and crops. The High Court had directed that any enquiry into the scope of patentability was not to be fettered by linguistic interpretation of what precisely could be a 'manner of manufacture' but rather should ascertain whether the claimed subject matter was a proper subject of letters patent under the Act. Emphasising the 'broad

---

[14] *Diamond v Chakrabarty* 100 S Ct 2204, 2208 (16 June 1980) (Being judicial exclusions that fall outside the scope of patent eligible subject in the United States. The eligibility inquiry under 35 USC § 101 provides that a claimed invention constitutes a new and useful subject matter as a process, machine, manufacture, or composition of matter, or any new and useful improvement thereof …).

[15] Patents Act 1977 (UK), s 1(2) and Article 52(2) of the European Patent Convention.

[16] See Australia Patents Act 1990 (Cth) s 18(1) and Patents Act 2013 (NZ) s 14 (referring to the 1623 (21 Jac 1, C3) Statute of Monopolies: 'An Act concerning monopolies and dispensations with penal laws and the forfeiture thereof.' Under the Statute of Monopolies, for an invention to constitute patent eligible subject matter it must be a 'manner of new manufacture' subject to the three main limitations on a grant of patent monopoly. Patents shall not be granted where a claim is contrary to the law, is mischievous to the State by raising prices of commodities at home or hurting trade, or is generally inconvenient.

[17] *National Research Development Corporation v Commissioner of Patents* (1959) 102 CLR 252.

[18] *Cancer Voices Australia v Myriad Genetics Inc* [2013] FCA 65 (15 February 2013) [110].

sweep' of the concept to encourage potential unforeseeable developments,[19] it was held that if a method 'has as its end result an artificial effect [or] artificially created state of affairs [and] possesses its own economic utility it will be patentable'.[20]

This direction continues to inform courts in the context of the need for a continual widening of the concept of an invention (which has been a characteristic of the growth of patent law) both in Australia[21] and the United States.[22]

Biotechnology patents have thus come a long way since the discovery of the deoxyribonucleic acid (DNA) sequence, genes and genetic engineering, to the 'recombinant DNA' (*r*DNA) technique allowing for genetically modified organisms and transgenic plants and animals. Prior to the *Myriad* debate, while the discovery of naturally occurring micro-organisms was not patentable, an isolated and purified gene or DNA molecule was a patentable invention. The United States Patent and Trademark Office (USPTO) had granted patents for isolated DNA molecules with the same sequence as a naturally occurring gene on the basis that the DNA molecule did not occur in nature in isolated form, as the process of synthesising or isolating it had changed its form.[23]

The United Kingdom, following the European Patent Convention (EPC), allows inventions of products consisting of or containing biological material and provides for isolated substances to be patentable even if these previously occurred in nature.[24] Thus 'an element isolated from the human body or otherwise by means of a technical process, including the sequence or partial sequence of a gene, may constitute a patentable invention, even if the structure of that element is identical to that of a natural element'.[25]

---

19 *National Research Development Corporation* (n 17), 269–71.

20 ibid 277.

21 *Apotex Pty Ltd v Sanofi-Aventis Australia Pty Ltd* [2013] HCA 50 (4 December 2013) [224] (Crennan and Kiefel JJ).

22 *Diamond v Chakrabarty* 100 S Ct 2204, 2205 (16 June 1980) ('Congress contemplated that the patent laws should be given wide scope, and the relevant legislative history also supports a broad construction').

23 *Association for Molecular Pathology v USPTO* 689 F 3d 1303, 1344 (Fed Cir. 16 August 2012).

24 Patents Act 1977 (UK), Schedule A2, Articles 1(a) & 2. See also Article 3(2) of the European Directive 98/44/EC on the legal protection of biotechnological inventions implemented by amendment to the UK Act which states: 'Biological material which is isolated from its natural environment or produced by means of a technical process may be the subject of an invention even if it previously occurred in nature'.

25 Patents Act 1977 (UK), Schedule A2, Article 5. See also Article 5(2) of the European Directive 98/44/EC on the legal protection of biotechnological inventions. See Nuffield Bioethics Report 2001 para 3.7 and para 3.18. Note, however, in *Kirin-Amgen Inc v Hoechst Marion Roussel Ltd* [2005] RPC 169: Lord Hoffman in the UK House of Lords drew a distinction between a discovery of nature's law and

Likewise, in New Zealand, the discovery of naturally occurring substances in nature is not patentable.[26] However, once isolated, purified or modified and removed from their natural state in any way to a degree that does not exist in nature (that is, an invention), a patent can be obtained. Thus patents for chemical substances that occur in nature are not necessarily excluded. A gene (DNA molecule) as a chemical substance when isolated and purified with the same sequence as a naturally occurring gene is considered patentable because it does not exist in nature in that form.[27]

The status quo, however, has been thrown into disarray by challenges to the patenting of genes and genetic material.

3. PATENT ELIGIBILITY OF BIOPRODUCTS – 'PRODUCTS OF NATURE' VERSUS 'PRODUCTS DERIVED FROM NATURE'

3.1 Genetically Modified Organisms

In *Diamond v Chakrabarty*[28] a patent was granted to Chakrabarty, who had discovered a process by which four different plasmids capable of degrading oil components could be transferred to a single bacterium. The new bacterial life form challenged the scope of patentable subject matter as to whether such a genetically modified organism was a mere discovery (that is, product of nature) or a man-made invention (that is, product 'derived' from nature).

The patent examiner had allowed the process claim for the method of producing the bacteria but rejected the claim for the bacteria on grounds that the micro-organisms were products of nature and as living things they did not constitute patentable subject matter.[29] The United States Supreme Court, however, cautioned against reading into the patent legislation limitations and conditions not expressed by the legislature, starting with the language of 35 USC § 101:[30]

---

the application of that discovery to state that DNA is not patentable. The claims in *Kirin-Amgen* related to DNA sequences were found to be invalid as natural occurring polypeptides (rEPO) that had been isolated before bearing the same chemical composition. Thus the isolated DNA was not new or patentable (at paras [5], [86]–[87], [96]–[101]).

[26] Patents Act 2013 (NZ) s 39(4) states: 'If a complete specification claims a new substance, the claim must not be construed as extending to that substance when found in nature'.

[27] Ministry of Economic Development Paper (NZ) *Implications of the Granting of Patents over Genetic Material*, paras 9 and 10. (File No. IP1.7.2.5)

[28] *Diamond v Chakrabarty* 100 S Ct 2204 (16 June 1980).

[29] ibid 2206.

[30] ibid 2207.

'Whoever invents or discovers any new and useful process, machine manufacture, or composition of matter, or any new and useful improvement thereof, may obtain a patent therefor, subject to the conditions and requirements of this title'.[31]

Delineating between products of nature and inventions, the Supreme Court held that Chakrabarty's oil-eating bacteria (this being a capability possessed by no naturally occurring bacteria) was a result of 'human ingenuity' with 'markedly different characteristics' and significant value for treating oil spills.[32]

3.2 *Oncomouse*

After *Chakrabarty*, a patent was next granted for the Harvard Oncomouse, a transgenic non-human mammal with an activated oncogene sequence introduced at embryonic stage to predispose the mice to cancer that was suitable for carcinogenic studies.[33] Led by the Supreme Court in *Chakrabarty*, the courts adopt in general that the types of patentable invention should be given 'wide scope' subject to the limitations spelt out in the legislation.[34] The complexity of genetic splicing was not part of the law of nature and this invention was held to be new, useful and non-obvious, as the Harvard researchers did not just find a 'naturally occurring oncomouse'.

As seen, the United States led the trend towards patentability of genetically modified organisms in the *Chakrabarty* and *Oncomouse* cases, with the United Kingdom and the European Union adopting a balancing exercise. In the European Union, a patent was initially declined for the Oncomouse under the EPC. Amid controversy and opposition from animal rights groups, it was finally granted on appeal in 1992, based on a utilitarian balancing exercise between the suffering of a transgenic animal and the usefulness of the invention.[35] Like *Chakrabarty*, however, new and non-obvious micro-organisms are patentable as well in the European Union under the EPC. New Zealand has

---

[31] As defined in 35 USC § 100(b) the word 'process' means process, art or method, and includes a new use of a known process, machine, manufacture, composition of matter or material.

[32] *Diamond* (n 28) 2208.

[33] USPTO Patent 4,736,866 (12 April 1988).

[34] *Diamond* (n 28) 2207.

[35] European patent 0 169762. Filed in 1985 but declined under Article 53(b) of the European Patent Convention. Patent granted on appeal amid controversy and 17 oppositions made by mainly animal welfare groups.

also allowed such inventions, with the new patent regime advocating a balancing exercise as well.[36]

The Supreme Court of Canada, however, went the other direction on the Oncomouse case.[37] The majority held that a genetically engineered non-human mammal was clearly not an 'art', 'process' or 'machine'. On whether a genetically altered higher life form (a conscious and sentient creature) could be a 'manufacture' or a 'composition of matter' under the provisions of the Canadian Statute,[38] the majority held that a composition of matter did not include such a higher life form. The majority asserted that the distinction between lower and higher life forms was defensible based on 'common sense' differences between the two even though it was not explicitly stated in the legislation.[39] The patent for the process by which the oncomice were produced was allowed, but the end products of the process (that is, the founder mouse and offspring) were not patentable.

The Supreme Court referred to the increasingly blurred lines in the new field of xenotransplantation with regard to human genes introduced into mammals such as a pig to make the animal's organs more acceptable to the human body for organ transplantation purposes, noting in this context the question raised by the Animal Alliance of Canada (as intervener) in their submission to the Court: 'The pig receives human genes. The human receives pig organs. Where does the pig end and the human begin?'[40]

3.3 Dolly the Sheep

Another landmark in genetic technology came with the arrival of Dolly, the first cloned sheep, in 2009. Scientists in the Roslin Institute, Edinburgh, were able to convert an adult sheep's cell into an embryo, which was then implanted into a surrogate mammal, where it developed into a baby animal and then the resulting new sheep. The patent claimed an invention for a method of cloning (by somatic cell nuclear transfer) and the production of transgenic animals as well as non-transgenics of high genetic merit.[41] The method was for reconstituting an animal embryo by cell fusion from a donor cell into a recipient cell that might then give rise to one or more animals. Dolly's birth, and the result-

---

36 Patents Act 2013 (NZ) s 15 (see examples listed).

37 *Harvard College v Canada (Commissioner of Patents)* [2002] 4 SCR 45 (Supreme Court of Canada 5 December 2002).

38 Patent Act RSC 1985, c P-4 (Canada).

39 *Harvard College* (n 37) [199].

40 ibid [180].

41 USPTO Patent 7514258 (7 April 2009).

ing ill health that forced her to be put down, prompted a fierce debate about the ethics and ramifications of cloning.

After obtaining the patent for the breakthrough on a method of cloning to create Dolly, the Roslin Institute later sought a patent for the product of that cloning method (that is, Dolly herself), or, as the claim read, for a 'live-born clone of a pre-existing, nonembryonic, donor mammal, wherein the mammal is selected from cattle, sheep, pigs, and goats'. The claim was denied by the United States Federal Circuit Court of Appeals.[42] Dolly the sheep was a product of nature or natural phenomenon and was therefore not patent eligible: 'Dolly herself is an exact genetic replica of another sheep and does not possess "markedly different characteristics from any [farm animals] found in nature".'[43]

The Court held that the claim had been anticipated and was obvious as the products were identical or substantially identical using the same cloning methods based on the prior art.

In summary, claims for 'process' patents by which bioproducts are produced appear to be less controversial. More concerning are 'product' patents such as the claims for the founder mouse and offspring and the cloned sheep that were rejected in the Canadian *Oncomouse* and United States *Roslin* cases. The decisions provide rationales for the exclusion of bioproducts from patentability on the grounds that a composition of matter does not include higher life forms, and that cloning of higher life forms relates to 'a product of nature'. In the following two sections, we focus on the continuing concerns expressed in the courts regarding product patents for isolated DNA sequences and examine how such concerns would pan out in the scenario of a 3D biofabricated product of a human organ.

4. ISOLATED DNA SEQUENCES – THE *MYRIAD* SAGA

4.1 United States: Judicial Voices from the District Court to the Supreme Court

The complexity of determining the threshold enquiry on patent eligibility was evident from the dissonant judicial voices in the controversial *Myriad Genetics* litigation that unfolded across the hierarchy of courts. This brought to the fore the discovery versus invention debate on isolated DNA sequences. The USPTO had granted patents for DNA sequences that were claimed in the

42 *In Re Roslin Inst.* (n 11).

43 ibid 1337 (citing *Diamond v Chakrabarty* 100 S Ct 2204, 2208 (1980)).

form of 'isolated DNA' on the premise that DNA should be treated no differently from any other chemical compound. Its purification from the body using well-known techniques transformed it into 'something distinctly different in character' and patentable. In response, a group of PhD researchers, medical organisations, counsellors and patients challenged the validity of the patents granted for the DNA and synthetically created DNA (cDNA) sequences predisposed to breast and ovarian cancer.

In the District Court, Sweet J disparaged the patentability of both the isolated DNA and cDNA sequences as they were not markedly different from native DNA, which was the physical embodiment of biological information (distinct in its essential characteristics from any other chemical found in nature).[44] He concluded that the DNA in an isolated form 'alters neither this fundamental quality of DNA as it exists in the body nor the information it encodes'.[45] The final isolated DNA in fact reached the same result as the DNA as it was in the body.[46] Likewise, the cDNA sequence was found in the human genome in the form of a naturally occurring pseudogene.[47]

The Federal Circuit Court of Appeals, on remand, however found both the isolated DNA and cDNA to be patent eligible subject matter.[48] The majority opinion, led by Lourie J, emphasised that an isolated DNA was just a fraction of the size of a naturally occurring DNA molecule and different as the covalent bonds in its backbone had been chemically severed.[49] They were 'markedly different' in their distinctive chemical structure and identity from native DNAs, which exist as one of 46 large, contiguous DNA molecules in the human body.[50] The isolated form was thus different from the natural DNAs,

---

44 *Association for Molecular Pathology v USPTO* 702 F Supp 2d 181, 185 (SDNY 29 March 2010).

45 ibid.

46 ibid.

47 ibid 230.

48 *Association for Molecular Pathology* (Fed Cir. 16 August 2012) (n 23) 1324. The question before the Court of Appeals was solely on the threshold test for patent eligibility in light of various Supreme Court holdings, particularly *Mayo Collaborative Services v Prometheus Laboratories, Inc* 132 S Ct 1289 (20 March 2012). In *Mayo* (a case demonstrating the 'law of nature' exclusion) the Supreme Court held that the patent was invalid as it claimed the underlying laws of nature and had no inventive step. The correlation of relationship between metabolite levels in the blood and likely harm or ineffectiveness of thiopurine drug dosage is a natural phenomenon, and merely 'applying' the above three steps is not sufficient to 'transform unpatentable law of nature' into an invention. 'The three steps add nothing specific to the laws of nature other than what is well-understood, routine, conventional activity, previously engaged in by those in the field.'

49 ibid 1328.

50 ibid 1328.

though ultimately derived from them. This reasoning that the isolated DNA molecules created in the laboratory were not found in nature and constituted a man-made product of human ingenuity resonated with the practice adopted by the USPTO.[51]

Moore J's concurring opinion, on the other hand, took on a different analysis without relying totally on Lourie J's conclusion. She upheld the validity of the claims to isolated DNA sequences based on the long-standing practice and guidelines of the USPTO and the 'settled expectations of the biotechnology industry'.[52] At the same time, though, Moore J had qualified that she might decide differently on the longer strands of isolated DNA containing most or all of a gene (as not patentable) were she to decide on a blank canvas.[53]

While Lourie and Moore JJ had focused on the differences between the isolated and naturally occurring DNAs, Bryson J, in his forceful dissent, (in part) focused on the similarity in structure as well as the utility between the isolated and native DNA. Finding nothing inventive in an isolation process that was a mainstay of a chemist's art,[54] he declined to give significant weight to the USPTO's practice on the premise that it lacked substantive lawmaking authority.[55] In his view, the information contained in the DNA sequences was the critical aspect of the molecules. However, he agreed that the cDNA was patent eligible as the end product was a human-made invention with a distinct structure (that is, introns had been removed from the cDNA segment and it now had additional utility).[56]

In a surprising twist to the saga that substantially undermined the USPTO's long-standing practice, the government as amicus curiae in the case did not defend the USPTO's position that isolated DNAs were patent eligible. It instead adopted a middle ground that the isolated and unmodified genomic DNAs were *not* patent eligible (that is, as a 'product of nature') since their nucleotide sequences existed because of evolution, not man. On the other hand, the cDNAs engineered by man were patentable.[57] The government's illustration featured an imaginary 'magic microscope'. If this imaginary microscope could zoom in onto the claimed DNA sequences as they were in the human body (which it could for the isolated BRCA1 and BRCA2 sequences) then it would not be patent eligible. If it could not (as in the case of the cDNAs

---

51 ibid 1332.
52 ibid 1344.
53 ibid 1343.
54 ibid 1355.
55 ibid 1357.
56 ibid 1356.
57 ibid 1326.

engineered by man to splice together non-contiguous coding sequences) then it was a patentable product 'derived from nature'.[58]

In June 2013, the United States Supreme Court held that 'genes and the information they encode are not patent eligible under § 101 simply because they have been isolated from the surrounding genetic material'.[59] Distinguishing *Myriad* from *Chakrabarty*, the Supreme Court held that an isolated DNA fragment was not patent-eligible subject matter. First, merely isolating the DNA from its surrounding genetic material using 'well understood' and 'widely used' processes did not significantly change it from its natural state. Second, the gene and the information encoded (without alteration) did not qualify it to be a product of human ingenuity.[60] Myriad Genetics made a discovery in uncovering the precise location and genetic sequences of BRCA1 and BRCA2, each a product of nature and not a genetically modified micro-organism. Such discovery, however brilliant, groundbreaking and useful, is not an invention under 35 USC § 101.[61] However, cDNA was deemed synthesised and artificial, and thus patentable.[62]

4.2 Australia: Arguments in the Federal Court, Full Federal Court and High Court of Australia

In the Federal Court of Australia at first instance, it fell to Nicholas J to decide in *Cancer Voices Australia v Myriad Genetics Inc*[63] whether the isolated DNA sequences were patent eligible subject matter. Nicholas J noted that the exclusion of products of nature argument did not accord with the principles of patent law as determined by Australia's highest court in the *NRDC* decision.[64] The seminal 'artificial state of affairs' requirement in *NRDC* was not directed towards a comparison of whether the composition of matter was a 'product of nature' or 'markedly different' under the United States *Chakrabarty* test.[65]

---

[58] ibid 1326.

[59] *Association for Molecular Pathology v Myriad Genetics Inc* 133 S Ct 2107, 2120 (13 June 2013).

[60] ibid 2119-2120 (13 June 2013). See also Judith L Toffenetti and Atabak R Royaee 'Patentability of 3D-printed organs' *Genetic Engineering & Biotechnology News* 15 May 2014 www.genengnews.com/insight-and-intelligence/patentability-of-3d-printed-organs/77900129/ accessed 3 December 2020.

[61] *Association for Molecular Pathology* (S Ct 13 June 2013) (n 59) 2117.

[62] ibid, at 2118–19.

[63] *Cancer Voices Australia v Myriad Genetics Inc* [2013] FCA 65 (15 February 2013).

[64] ibid [81].

[65] ibid [101] and [103].

Native DNA does not exist outside the cell and isolated DNA does not exist inside the cell. Thus, the isolated DNA was a product of human intervention.[66] Providing the reasons for his finding that isolated nucleic acid sequences were an 'artificial state of affairs', Nicholas J stated that the scope of the concept of 'manner of manufacture' was intended to be broad and inclusive.[67] The extraction and purification steps that resulted in the isolated nucleic acid sequences required intensive research and immense intellectual effort, and it would be a mistake and inconsistent with the purposes of the legislation not to reward such efforts in accordance with the broad language of the *NRDC* decision.[68]

On appeal, in *D'Arcy v Myriad Genetics Inc*[69] the Full Federal Court of Australia, agreeing with the court below, likewise applied the law as explained in *NRDC*. Interestingly, by then the Full Court had the benefit of the United States Supreme Court determination that isolating the DNAs was mere discovery, however brilliant. As observed, the United States Supreme Court had focused on the genetic information encoded in the genes excluding it from patentability.[70] However, the Full Court remained unpersuaded by the United States Supreme Court's 'product of nature' conclusion, finding the laws of nature argument not quite a useful tool of analysis.[71]

The Full Federal Court of Australia found it more justifiable to view the claimed subject matter as a chemical compound.[72] It favoured the line of reasoning adopted by the United States Federal Circuit Court of Appeals which focused on the claimed composition and the product as it was in nature to discern whether the alteration by human intervention had resulted in 'markedly different' or 'distinctive characteristics'. The Full Court found this reasoning similar to the *NRDC* requirement.[73] Focusing, likewise, on the differences in structure and function (rather than similarities) caused by human intervention, the Full Court concluded that isolated DNA and RNA resulted in an artificially

---

[66] ibid [108].

[67] ibid [109] and [123].

[68] ibid [109].

[69] *D'Arcy v Myriad Genetics Inc* [2014] FCAFC 115 (7, 8 August 2013, 5 September 2014).

[70] ibid [129] and [138].

[71] ibid [13]. Note also the IP Final Report, Centre for International Economics, Australia, May 2013 'Economic Analysis of the Impact of Isolated Human Gene Patents' p23 (judicial interpretation recognised that mere discoveries, ideas, scientific theories and laws of nature do not satisfy the test for a manner of manufacture and invention).

[72] ibid [143].

[73] ibid [142].

created state of affairs of economic benefit and was thus properly the subject of letters patent for an invention under the Australian Act.[74]

On final appeal to the High Court of Australia, the principles developed for the application of s 6 of the *Statute of Monopolies* as framed by them in *NRDC* were reiterated.[75] The High Court acknowledged that in determining a 'manner of manufacture', case law had developed consistently and broadly in line with the development of patent law.[76] However, Gordon J said that Myriad's claims were to a product, that is, to 'every isolated example of the BRCA1 gene, or a portion of the BRCA1 gene, which discloses the existence of one or more specific mutations or polymorphisms'. She thus concluded that the BRCA1 gene was not a proper subject for the grant of a patent under the Australian Act.[77] Gordon J said that while Myriad identified the location of the BRCA1 gene, it did not 'create, make or alter any of the nuclei acid sequence' in the gene.[78] In fact, her conclusion that even cDNA should not be patentable went further than the stance taken in the United States Supreme Court.[79]

The *Myriad* cases in the United States and Australia show a trend towards narrowing down the patentability of isolated genes. However, the divergence regarding the patentability of cDNA is worth noting. While the United States *Myriad* case concluded that cDNA was patent eligible as the end product was a human-made invention with a distinct structure, the Australian High Court in *D'Arcy v Myriad* considered otherwise.

Dreyfuss and colleagues argue that the Australian approach was more nuanced by adopting a 'factorial test' in which a non-exhaustive list of factors was taken into deliberation.[80] These were mostly external factors that needed to be considered while determining patentability: the purpose of patent legislation, the potential negative or chilling effects on innovation and other activities, an assessment of conflicting public and private interests, whether according patentability would enhance or detract from the coherence of the law, a State's international obligations, and finally, whether it would involve a kind of law making that should be done by legislature.[81] The plurality factorial test would provide a holistic lens for the social function of patents and could be used in interpreting patentability while engaging a broader societal agenda.

---

[74] ibid [155].

[75] *D'Arcy v Myriad Genetics Inc* [2015] HCA 35 (7 October 2015) [18].

[76] ibid (citing *National Research Development Corporation* (n 17) 270).

[77] ibid [177]–[178].

[78] ibid [248].

[79] ibid [283].

[80] RC Dreyfuss, J Nielsen and D Nicol 'Patenting nature – a comparative perspective' (2018) Journal of Law and Biosciences 550–89 doi:10.1093/jlb/lsy021.

[81] ibid 570–1.

5. PATENTABILITY SCENARIO FOR 3D BIOFABRICATED PRODUCTS (HUMAN ORGANS)

5.1 Biofabricated Human Organs

Although the reality of functional biofabricated human tissues and organs may not be viable for a while yet, breakthroughs in this area have resulted in applications to the USPTO for methods of fabricating interconnected model vasculature based on bioprinting techniques.[82] As one barrier to the creation of large, whole organ replacement relates mainly to nutrient and waste transport in the body, a main goal is to be able to make whole organ replacements compatible with the relevant blood vessel network in the human body. Other breakthroughs to overcome key challenges in this area have enabled scientists and researchers to develop platforms for living 3D biofabricated tissues or organs.[83] Patents have been granted for self-assembling cell aggregates for 3D tissue and organs. Besides the method involved, the invention includes a 3D layered structure, though the assembly of vascularised 3D soft organs has not yet been accomplished.[84] Promising computer-aided technologies have enabled organ printing aimed at eventually constructing functional organs layer by layer.[85] Likewise, a number of approaches have been described for either in vitro or in vivo growth of cells in 3D tissues or organs while minimising cell damage.[86]

---

[82] United States Patent Application (8 March 2012).

[83] United States Patent Application Nos 20130164339 (27 June 2013) and 20140099709 (10 April 2014). See also LE Bertassoni et al 'Hydrogel bioprinted microchannel networks for vascularization of tissue engineering constructs' Lab Chip (7 July 2014) 14(13):2202–11.

[84] USPTO Patent 8,241,905 (14 August 2012) (self-assembling cell aggregates for 3D tissue and organs).

[85] Ibid. See also Edison Bicudo, Alex Faulkner and Phoebe Li (2020) 'Software, risks, and liability: ongoing and emerging issues in 3D bioprinting' Journal of Risk Research. DOI: 10.1080/13669877.2020.1848904.

[86] US Patent Application No. 20140099709 (living 3D tissue and organs). James O'Toole '3-D-printed organs are on the way' *CNNMoney* (New York) 4 November 2014 https://money.cnn.com/2014/11/04/technology/innovationnation/3d-printed-organs/ accessed 3 December 2020.

The major players in this area of biofabrication are the United States,[87] the United Kingdom and the European Union, as well as Australia[88] and China. The Chinese government opened its first 'Medical 3D Printing and Research and Development Centre' in Changchun to construct the world's first 3D printed elbow replacement surgery.[89]

5.2 United States and Australia

With the final stance on the non-patentability of isolated DNAs drawn in the highest courts in the United States and Australia, would a biofabricated organ be able to surmount the patent eligibility hurdle as a product of human ingenuity under the second prong of the *Chakrabarty* test, subject to the further requirements for novelty, non-obviousness and utility?[90] Proponents of patents for such inventions argue that though a biofabricated organ replicates the design, shape and function of a human organ, its composition or manufacture nevertheless is a product of human ingenuity requiring highly scientific methods and precise machinery, separating it from a product of nature.[91] However, as the United States Supreme Court had pointed out, extensive

---

87 Ben Coxworth 'Organovo now selling tiny 3D-printed human livers' *New Atlas* 19 November 2014 www.gizmag.com/organovo-exvive3d-liver-models/34843/ accessed 4 December 2020. Organovo has been researching and creating miniature 3D bioprinted human livers, called exVive3D, for experimental purposes from a 3D bioprinter. The miniature livers are made up of living human liver tissue producing liver proteins such as albumin, fibrinogen and transferrin like the natural organ and can synthesise cholesterol.

88 See 'Australia develops bio 3D printing for injury treatment' 17 February 2015 www.3d-printers.com.au/2015/02/17/australia-develops-bio-3d-printing-for-injury-treatment/ accessed 23 September 2019. (The Queensland University of Technology in Australia had been working on the biofabrication process in 3D printing which involved adding a biological/living complexity to a rapid prototyping technique. Though there were obstacles to overcome in terms of sterility, costs and approval there had been tremendous progress in printing cells within hydrogels for cartilage repair and 3D printed scaffolds for cranial application.) The Australian Intellectual Property Report 2014 reported that the University of Woolongong used 3D printers to bio-fabricate skin, cartilage, arteries and heart valves, while scientists at Princeton University had achieved 3D bioprinting of organ tissues in the form of blood vessels and a 'functional ear'.

89 See 'China Starts up The Medical 3D Printing R&D Centre' 3D print Today 5 May www.3d-print.today/China_Starts_Up_The_Medical_3D_Printing_RD_Centre accessed 4 December 2020.

90 See Toffenetti and Royaee 'Patentability of 3D-printed organs' (n 60).

91 Martin and Horton 'Patent eligibility of 3D printed organs will soon be an issue' (n 6).

research effort alone is insufficient to satisfy patent eligibility.[92] While many inventive steps are needed to extract the human cells and grow the human tissue via a bioprinter (not unlike the extracting of native DNA from its cellular environment), it is undisputed that the biofabricated organ is of the same composition as the original.

First, to be a patentable bioproduct of 'human ingenuity derived from nature' under the Chakrabarty test, the biofabricated organ must show 'markedly different characteristics'. The test necessitates something more, as in a 'transformation' of a new and different article to emerge with a distinctive character or use.[93] We would recall Sweet J's argument, in the United States District Court in *Myriad*, that isolated DNA 'alters neither this fundamental quality of DNA as it exists in the body nor the information it encodes'.[94] On a similar note, a biofabricated organ is akin to a 'product of nature' comprising the same human cells and proteins as the organs in the body.

Second, the dissent in the Federal Circuit Court of Appeals in *Myriad* had attempted to draw an analogy between the isolated DNAs and a kidney removed from the human body to demonstrate why both would not be patent eligible subject matter. The analogy was rejected by the majority on the basis that a kidney is an organ, not a well-defined composition of matter or an article of manufacture specified by § 101.[95] Isolating or extracting a kidney from a body does not result in a patent-eligible composition or require extensive research to locate, identify and isolate, as does a gene.[96] Thus, if it is to be understood that an isolated body organ would not be properly characterised as a composition of matter for patentability, it follows that neither would a 3D biofabricated organ replicated from extracted stem cells of the natural human organ be patentable.

Third, as noted above, the Supreme Court of Canada held that the Oncomouse was not an art, process or machine and that the scope for patentability does not include a higher life form. In the United States Federal Circuit Court of Appeals, the product claim on the cloned sheep, Dolly, was rejected as she was an exact genetic replica of another sheep with no 'markedly different characteristics' from nature's farm animals. The claim was not new (and had been anticipated) and was obvious.[97] A parallel argument could then prevail

---

92 *Association for Molecular Pathology* (S Ct 13 June 2013) (n 59) 2118 (13 June 2013).

93 *Association for Molecular Pathology* (SDNY 29 March 2010) (n 44) 222.

94 ibid 185.

95 *Association for Molecular Pathology* (Fed Cir. 16 August 2012) (n 23) 1332.

96 ibid 1332.

97 *In Re Roslin Inst.* (n 11) 1337 (citing *Diamond v Chakrabarty* 100 S Ct 2204, 2208 (1980)).

that a biofabricated kidney as an exact replica of a person's kidney is a product of nature. Such a claim (as asserted in *Dolly*) might well be anticipated and obvious as the products are identical or substantially so based on the prior art.

5.3 United Kingdom and New Zealand

The patentability of 3D biofabricated products and processes has been examined specifically through the *morality and ordre public* test and the medical treatment exception in regenerative medicine and tissue engineering in an earlier article by one of the authors of this chapter.[98] 3D biofabricated human organs may be excluded if it is a claim for a method of treatment by surgery, diagnosis or therapy in the United Kingdom (and New Zealand), where legislation provides for an explicit medical treatment exception.[99] However, if a method claim is construed under the non-surgical aspect of making and printing of a 3D biofabricated human organ (in contrast with the surgical implantation of a 3D biofabricated organ) it could be patentable and not fall under the medical treatment exclusion.[100]

The EU Biotech Directive elaborates the principles for excluding patentability in the EPC and specifies that 'processes for cloning human beings' shall be unpatentable due to the commercialisation of which is considered contrary to *ordre public* or morality. It makes no reference to the products of human cloning or parts of human cloning. Hence it is less clear whether the case of biofabricated products may be caught by the human 'cloning' exclusion.[101] Further, the EU Biotech Directive expresses that 'Biological material which is isolated from its natural environment or produced by means of a technical process may be the subject of an invention even if it previously occurred in nature'.[102]

In New Zealand (guided too by the Australian *NRDC* decision), the very act of removal of the cells and bioprinting an organ would amount to 'an artificial state of affairs' even if the physical properties of the product are the same. As Nicholas J earlier noted, claims for 'tangible materials' such as a genomic DNA sequence do not exist outside the cell and in that sense it can be said that

---

98 Li '3D Bioprinting Technologies: Patents, Innovation and Access' (n 13).

99 Patents Act 1977 (UK) s 4A(1); Patents Act 2013 (NZ) s 16.

100 Li '3D Bioprinting Technologies: Patents, Innovation and Access' (n 13).

101 See Art 53(a) EPC; Art 6 Directive 98/44/EC of the European Parliament and of the Council of 6 July 1998 on the legal protection of biotechnological inventions (Biotech Directive).

102 Art 3(2) Biotech Directive.

an isolated chemical is not the same as the chemical *in situ*.[103] Furthermore, the isolated chemical can be manipulated and utilised in ways that the chemical in situ may not be.[104] A biofabricated organ based on such an observation would be able to achieve a separate and additional result possessing its own economic utility.

Thus, the United Kingdom (guided by the EU Biotech Directive) and New Zealand allow for a different (and more expansive) test. However, while it appears that the method itself may be able to avoid the medical exception, the question as to whether the end product (a biofabricated organ) is patent eligible matter remains open to interpretation by the courts.

6. CONCLUSION

We have examined the relevant precedents on patenting biomaterials in comparative jurisdictions, which demonstrate a varying degree of regulatory discretion and highlight that there may not exist a 'one-size-fits-all' answer to patenting biofabricated products. Jurisdictions may opt for avenues specifically tailored to the development of domestic biotechnology industry as well as the balance of access to healthcare delivery. Particularly, where there is scope for public policy, the *morality* and *ordre public* considerations may be engaged as a means to embracing public engagement in future patent policymaking.

The highly controversial area of genetic engineering and patents on DNA sequences raises moral and ethical issues, prompting a Circuit Court of Appeals judge in the United States to comment: 'It is tempting to use our judicial power in this fashion, especially when the patents [...] raise substantial moral and ethical issues [...] awarding a property right to isolated portions of human DNA – the very thing that makes us humans, and not chimpanzees.'[105]

The outpouring of scientific activity and advances in biomedical research, pharmaceuticals and access to health require incentives and encouragement. However, as Bryson J pointed out in his dissent in *Myriad*, while incentive is crucial to encourage and reward research in the difficult and important field of biotechnology, there is another side to the coin. Broad claims to genetic material pose significant obstacles to future research that needs to use such materials as building blocks. However, firms developing those technologies

[103] *D'Arcy* [2014] FCAFC 115 (n 69) [191] (referring to *Cancer Voices* [2013] FCA 65 (n 63) [76].

[104] ibid.

[105] *Association for Molecular Pathology* (Fed Cir. 16 August 2012) (n 23) 1346 (per Moore J).

are encountering a thicket of patents and require the licensing of thousands of these from many different parties and at prohibitive cost.[106]

With international initiatives to control illegal trafficking in organs and public outcry at such atrocities,[107] the World Health Organization has taken cognisance of the state of affairs caused by the organ shortage. It has prompted many countries to develop procedures and systems to increase supply. However, this may well have resulted instead in stimulating commercial traffic in human organs, particularly from living donors who are unrelated to recipients, and from medical centres advertising their ability to perform transplants and to supply donor organs for a single, inclusive charge.[108] Travel for such purposes to countries where prohibitions on commercialisation are not enforced is rife in the organ trade.

Biofabrication has the potential to solve one of the problems related to access to medicines. The viability of personalised medicine inventions[109] through bioprinting and matching of cell type reduces rejection rates and clinical testing of drugs and alleviates the ethical dilemma involved with of the practice of depending on executed prisoners' human organs. Thus, high obstacles are needed to justify the protection of an exclusive monopoly under the patent system, to ensure that intellectual property rights do not become a barrier to access to medical technology and health care in the biofabrication field. More so, we dread to envisage a world in which biofabricated organs prolong lives at the expense of owning our body parts.

---

[106] ibid 1357 (Bryson J dissenting – citing Secretary's Advisory Comm. on Genetics, Health, and Society, Dep't of Health & Human Servs, *Gene Patents and Licensing Practices and Their Impact on Patient Access to Genetic Tests* 49–62 (2010)).

[107] These include: United Nations Human Rights 'Protocol to Prevent, Suppress and Punish Trafficking in Persons' 15 November 2000 (Article 3 includes 'removal of organs' and its subsequent sale as an end purpose of trafficking); World Health Organization 'WHO Guiding Principles on Human Cell, Tissue and Organ Transplantation' (as endorsed by the sixty-third World Health Assembly in May 2010, in Resolution WHA63.22); Council of Europe Convention against Trafficking in Human Organs *CETS 216 – Trafficking in Human Organs* 25.III.2015 (which would make it a crime to extract organs from people without their 'free, informed and specific consent').

[108] 'WHO Guiding Principles on Human Cell, Tissue and Organ Transplantation' (n 107) para 2 www.who.int/transplantation/Guiding_PrinciplesTransplantation_WHA63.22en.pdf accessed 4 December 2020.

[109] Robin Feldman 'Whose body is it anyway? Human cells and the strange effects of property & intellectual property law' (2011) 63 Stan L Rev 1377, 1386.

5. A problem of interpretation: is Article 53(a) EPC necessarily restrictive of emerging technologies?

Ella O'Sullivan

1. INTRODUCTION

The engagement with *ordre public* and morality in the European Patent Office (EPO) has been described as 'incoherent, unduly restrictive and blind to the regulatory challenges presented by emerging technologies in the current social and political context'.[1] This seems to stem in part from a reluctance to engage with *ordre public* and morality in fear that the EPO might act as a 'moral censor' in the invocation of Article 53(a) EPC.[2] Therefore, in the long-running *Harvard Oncomouse* application, where the provision was initially viewed as irrelevant,[3] an evolving and ultimately confusing approach to *ordre public* and morality emerged, culminating in two balancing tests and a general *ordre public* and morality assessment.[4] Inconsistent and incoherent approaches to *ordre public* and morality result in confusion for patent applicants, inventors and potential opponents, particularly in the arena of emerging technologies, where such concerns are more likely to arise.[5] It further results in an isolation-

[1] J Pila, 'Adapting the *ordre public* and morality exclusion of European patent law to accommodate emerging technologies' 38 (2020) Nature Biotechnology 555, 555.

[2] *R v Leland Stanford/ Modified Animal* [2002] 2 EPOR 16, Reasons for the Decision [8].

[3] *Transgenic Animals/Harvard* Decision of the Examining Division (1989) OJ EPO 451, Reason for the Decision, [10.3]. This was in contrast to the preliminary decision of Primary Examiner Gugerell (who also sat on the Examining Division). There is an earlier case, *Antihistamines/EISA* T_0164/83 OJ EPO (1987) 149, noted by Moufang, where there was a question of animal suffering and no reference to Article 53(a) was made. See R Moufang, 'Patenting of human genes, cells and parts of the body? The ethical dimensions of patent law' (1994) 25 IIC 487, 500.

[4] T0315/03 *Transgenic Animals/Harvard* (2005) OJ EPO 246/ (2006) OJ EPO 15.

[5] J Pila, 'Adapting the *ordre public* and morality exclusion of European patent law to accommodate emerging technologies' 38 (2020) Nature Biotechnology 555 556.

ist patent system which may ultimately fail the society that should benefit from new emerging technologies.[6]

If the patent system exists within a framework of European and EU fundamental rights then it is not tenable to remove the *ordre public* and morality exclusion from the domain of patent law.[7] Indeed, it is requisite that these concepts are treated with primary importance in the evaluation of patent applications.[8] This will only work if the boards of the EPO adopt a consistent and coherent approach to the exclusion. Unfortunately, this is not presently the case. This chapter will examine how the inconsistent and variable interpretations of the Article 53(a) EPC provision present a greater obstacle to the patenting of emerging technologies than the provision itself.

Article 53(a) EPC sets out that patents will not be granted for 'inventions the commercial exploitation of which would be contrary to 'ordre public' or morality, provided that the exploitation shall not be deemed to be so contrary merely because it is prohibited by law or regulation in some or all of the Contacting States'. This chapter will examine the problematic interpretation of '*ordre public* or morality' and the standards and tests utilised to evaluate whether the potential commercial exploitation of the invention would contravene these principles. The imprecise focus of the provision on commercial exploitation will be considered, and finally the utility of the express exclusions introduced by Article 6(2) of Directive 98/44/EC on the legal protection of biotechnological inventions (Biotech Directive) and incorporated as Rule 28(1) EPC will be assessed.[9]

---

[6] The broader social context of patent law is highlighted by Walsh and Hawkins. K Walsh and N Hawkins, 'Expanding the role of morality and public policy in European patent law' in P Torremans (ed), *Intellectual Property Law and Human Rights* (4th edition, Wolters Kluwer, 2020) 4 & 6 (online) available at http://hdl.handle.net/10871/40561 last accessed 28 March 2021. See also J Pila, 'Adapting the *ordre public* and morality exclusion of European patent law to accommodate emerging technologies' 38 (2020) Nature Biotechnology 555, 557. McMahon highlights the potential downstream problems when broader social concerns are not sufficiently engaged with in the patent system. See A McMahon, 'Biotechnology, health and patents as private governance tools: the good, the bad and the potential for ugly?' 3 (2020) IPQ 161.

[7] Directive 98/44/EC on the legal protection of biotechnological inventions (1998) OJ L 213/13 Recital 16; C-34/10 *Brüstle v Greenpeace eV* ECLI:EU:C:2011:669 [32].

[8] D Beyleveld, R Brownsword and M Llewelyn, 'The morality clause of the Directive on the Legal Protection of Biotechnological Inventions: conflict, compromise and the patent community' in R Goldberg and J Lonbay (eds), *Pharmaceutical Medicine, Biotechnology and European Law* (Cambridge University Press, 2001) 157, 161; M O'Sullivan, *Biotechnology, Patents and Morality: A Deliberative and Participatory Paradigm for Reform* (Routledge, 2020) 74.

[9] Directive 98/44/EC on the legal protection of biotechnological inventions (1998) OJ L 213/13.

2. *ORDRE PUBLIC* OR MORALITY

The first problem with the application of Article 53(a) EPC is the use of contrasting approaches to the key elements of *ordre public* and morality. This is apparent both in their individual interpretations and in how they interact with each other. *Ordre public* does not have a direct translation into English but as a concept has been described as denoting fundamental principles of law,[10] 'the structure of social relations governed by the Rule of Law',[11] or some form of 'fundamental value system'.[12] The focus on fundamental values is apparent in the T0149/11 *Slaughterhouse* decision whereby it was set out that *ordre public* should be 'defined by norms that safeguard fundamental values and rights such as the inviolability of human dignity and the right of life and physical integrity'.[13]

The concept of *ordre public* was considered to be of such importance in *Slaughterhouse* that the Technical Board of Appeal was willing to refuse patent protection notwithstanding its view that Article 53(a) EPC was 'neither intended nor suitable for effectively restricting' inventions that fell within its remit.[14] In this context, the overall importance of the exclusion rested on the need of the patent system to maintain public trust: 'This public trust would erode if the broader public outside of the patent profession would perceive that

---

[10] See for example, G_0002/06 Comments by the President of the EPO (Incoming letter in referral procedure 28/09/06) 21 and 51 available in All Documents for EP96903521 at https://register.epo.org/espacenet/regviewer last accessed 28 March 2021. All relevant documents (including unreported material) may be found by searching the patent or application number in the EPO Register. J Straus, 'Patenting human genes in Europe – Past Developments and Prospects for the Future' (1995) 26 IIC 920, 930 and 932; U Schatz, 'Patentability of genetic engineering inventions in European Patent Office practice' (1998) 29 IIC 2, 14; R Moufang, 'The concept of '*ordre public*' and morality in patent law' in G Van Overwalle (ed), *Patent Law, Ethics and Biotechnology* (Katholieke Universiteit, Brussels, 1998) 68, 71; R Moufang, 'Patenting of human genes, cells and parts of the body? The ethical dimensions of patent law' (1994) 25 IIC 487, 503.

[11] D Beyleveld and R Brownsword, *Mice, Morality and Patents* (London: Common Law Institute of Intellectual Property, 1993), 62.

[12] A Warren Jones, 'Finding a "common morality codex" for biotech – a question of substance' (2008) 39 IIC 638, 651.

[13] T0149/11 *Method and device for processing a slaughtered animal or part thereof in a slaughterhouse* EP819381 (24 January 2013), Reasons for the Decision [2.5] available at www.epo.org/law-practice/case-law-appeals/recent/t110149eu1.html last accessed 28 March 2021.

[14] T0149/11 *Method and device for processing a slaughtered animal or part thereof in a slaughterhouse* EP819381 (24 January 2013), Reasons for the Decision [2.6].

a morally unacceptable condition – here the ownership of a human being – was endorsed by the patent system.'[15]

However, on other occasions the role of *ordre public* has focused on more everyday, specific concerns. In T0356/93 *Plant Genetic Systems* it was considered to encompass 'the protection of public security and the physical integrity of individuals as part of society' as well as 'the protection of the environment'.[16] These elements might be seen to represent practical manifestations of fundamental principles of law or society, although this context is not indicated in this decision. In the later case of T0866/01 *Euthanasia Compositions*, issues of public safety and the environment were indeed seen as flowing from fundamental social and legal norms.[17] Notwithstanding this, in most decisions the concept of *ordre public* (or morality) is not defined at all. For instance, in the recent T1553/15 *Rabbit Skin Extract/VanWorld* case no definition of *ordre public* or morality was propounded.[18]

The companion morality element is defined as relating to ethical values of society[19] or law[20] or, as per Beyleveld and Brownsword's construction, 'any

---

[15] T0149/11 *Method and device for processing a slaughtered animal or part thereof in a slaughterhouse* EP819381 (24 January 2013), Reasons for the Decision [2.6]. This case, which takes an unusually strong approach to *ordre public* and morality, rejected an application whereby the drafted invention included the role of a human 'observer' within its scope. This was rejected as it could be read to include a human as part of the invention, thus contravening fundamental human rights.

[16] T0356/93 *Plant Genetic Systems* (1995) OJ EPO 545, Reasons for the Decision [5]. This definition in T0356/93 *Plant Genetic Systems* is criticized by Pila as being unduly narrow. J Pila, 'Adapting the *ordre public* and morality exclusion of European patent law to accommodate emerging technologies' 38 (2020) Nature Biotechnology 555, 555.

[17] T0866/01 *Euthanasia Compositions/Michigan State University* EP (11 May 2005), Reasons for the Decision, [6.9] available at www.epo.org/law-practice/case-law-appeals/recent/t010866eu1.html last accessed 28 March 2021.

[18] T1553/15 *Rabbit Skin Extract/VanWorld* EP1557171 (28 September 2020) available at www.epo.org/law-practice/case-law-appeals/recent/t151553eu1.html last accessed 28 March 2021.

[19] See for example R Moufang, 'The concept of '*ordre public*' and morality in patent law' in G Van Overwalle (ed), *Patent Law, Ethics and Biotechnology* (Brussels: Katholieke Universiteit, 1998) 68, 71; R Moufang, 'Patenting of human genes, cells and parts of the body? The ethical dimensions of patent law' (1994) 25 IIC 487, 503; U Schatz, 'Patentability of genetic engineering inventions in European Patent Office practice' (1998) 29 IIC 2, 14; O Mills, 'Biotechnology and the ethical and moral concerns of European patent law' (2000) 6 *Bar Review* 46, 48; G_0002/06 Comments by the President of the EPO (28/09/06), 21 and 51; T_0356/93 *Plant Genetic Systems* (1995) OJ EPO 545, Reasons for the Decision [6]; T0866/01 *Euthanasia Compositions/Michigan State University* (11 May 2005), Reasons for the Decision [6.12].

[20] J Straus, 'Patenting human genes in Europe – past developments and prospects for the future' (1995) 26 IIC 920, 932.

matters of public morality that are not directly implicated within the idea of the Rule of Law'.[21] This would seem to indicate some crossover between the two concepts.[22] The definition of morality provided by the Technical Board of Appeal in *Euthanasia Compositions* is suggestive of overlap in describing morality as 'fundamental' and 'extra-legal' yet underplays its importance by denoting it as something already decided by European states:

> Morality is one of the fundamentals of our legal system and at the same time forms the basis for the inclusion of extra-legal principles of ethics in the law. The legal approach based on morality for the EPC can be found in the concepts of the European cultural and legal systems. Morality constitutes actual ethically-based norms of behaviour that have become socially binding through being generally accepted.[23]

This echoes the earlier definition in *Plant Genetic Systems* whereby morality is set out as 'related to the belief that some behaviour is right and acceptable whereas other behaviour is wrong, this belief being founded on the totality of the accepted norms which are deeply rooted in a particular [here European] culture'.[24]

The Technical Board of Appeal in *Euthanasia Compositions* viewed morality as not something to be determined by the EPO but rather something determined elsewhere.[25] The general description of *ordre public* and morality

---

[21] D Beyleveld and R Brownsword, *Mice, Morality and Patents* (London: Common Law Institute of Intellectual Property, 1993) 62. To demonstrate what is meant by the Rule of Law, Beyleveld and Brownsword cite J Bell at footnote 83: '[T]he rule of law is served in that the judge is required to make specific legal decisions fit coherently with the underlying purposes of the legal system. It may be that the various values expressed in doctrines of public policy are more ideal than actual in the current state of the law, as are ideas of integrity and coherence. But such an ideal aspect to law is important as a regulative device within the institution of law. It encourages decision-makers to make the law more fully fit for the ideal.' They further expand on this line of reasoning by describing 'ordre public' as constituting 'the minimum conditions for regulation in accordance with the Rule of Law' (77).

[22] Beyleveld and Brownsword suggest that '"[o]rdre public" and "morality" both belong within the category of "public morality". There is, in other words, a sense in which "ordre public" is a part of "morality".' D Beyleveld and R Brownsword, *Mice, Morality and Patents* (London: Common Law Institute of Intellectual Property, 1993) 78, fn 99.

[23] T0866/01 *Euthanasia Compositions/Michigan State University* (11 May 2005), Reasons for the Decision [6.12].

[24] T0356/93 *Plant Genetic Systems* (1995) OJ EPO 545, Reasons for the Decision [6].

[25] T0866/01 *Euthanasia Compositions/Michigan State University* (11 May 2005), Reasons for the Decision [6.12].

in Recital 39 of the Biotech Directive as 'correspond[ing] to ethical or moral principles in a Member State', albeit at first instance directed at domestic laws, further compounds the interpretation of *ordre public* and morality in the context of the European patent. These references to recognised principles undermine the provision, as consensus on emerging technologies rarely exists, and may indirectly represent another unsuccessful attempt to 'narrow' the scope of Article 53(a) EPC in the EPO.[26]

The effort to find a definitive interpretation of *ordre public* and morality is undermined by the uncertainty as to whether these two terms may be considered as conjunctive or disjunctive concepts (or some combination of the two). As evidenced above, Recital 39 of the Biotech Directive defines them collectively. This additional problem of interpretation and application is apparent in T0315/03 *Harvard Oncomouse*, where the Technical Board of Appeal acknowledged that both concepts 'may form the basis of two separate objections either or both of which can be raised in a particular case'.[27] In this instance, the provision was applied using a balancing test which did not distinguish between *ordre public* and morality factors.[28]

This failure to definitively distinguish between the two, or, alternatively, precisely construe the two as an overarching concept, opens the door for the inclusion of a vast array of arguments and evidence in opposition procedures, which in turn allows the boards of the EPO to reject these considerations as irrelevant.[29] It results in potentially unnecessary litigation and leaves both opponents and patent applicants unsure of what might engage Article 53(a) EPC in practice. The best solution would be for the boards of the EPO to adopt a consistent approach to the interpretation of *ordre public* and/or morality. Such a definition would need to be necessarily general but it would resolve the current confusion about the precise definitional limits of the core Article 53(a) EPC provision. Ideally, this would require the legislature to undertake an analysis of what is meant by the terms at first instance. To maintain reasonable flexibility to deal with unforeseen problems with emerging technology,

---

[26] K Walsh and N Hawkins, 'Expanding the role of morality and public policy in European patent law' in P Torremans (ed), *Intellectual Property Law and Human Rights* (4th edition, Wolters Kluwer, 2020) 10 (online) available at http://hdl.handle.net/10871/40561 last accessed 28 March 2021.

[27] (2005) OJ EPO 246/ (2006) OJ EPO 15, Reasons for the Decision [10.2] and [4.2].

[28] (2005) OJ EPO 246/ (2006) OJ EPO 15, Reasons for the Decision [13]–[13.2.21].

[29] See for example the Opposition Division in *Howard Florey/Relaxin*, where arguments related to slavery submitted by the opponents were deemed to constitute 'a fundamental misunderstanding of the effects of a patent'. (1995) OJ EPO 388, Reasons for the Decision 6.3.3.

the determination of a non-exhaustive list of factors which would trigger an Article 53(a) EPC assessment might be of assistance in this regard.[30]

Complicating the issue further is the absence of consensus regarding what standard determines if a patent must be rejected under Article 53(a) EPC. It is therefore unclear to what degree issues of *ordre public* and/or morality must prove problematic in order to merit patent rejection. Similarly, there is confusion regarding the method of determining if any such variable standard has been met.

3. STANDARDS AND METHODOLOGIES

3.1 Standards

It is possible to glean four key standards expressed by the boards of the EPO in the application of Article 53(a) EPC: the unacceptability standard;[31] the stand-

[30] See for instance the Irish Domestic Violence Act 2018, whereby section 5 directs the relevant court to consider various elements to determine if a domestic violence order should be granted while allowing the court to also take into account any other relevant factors at its discretion. This ensures that unexpected behaviour that comes within the scope of domestic violence is not unduly excluded from the scope of the legislation. Pila envisages a triage screening process specifically for emerging technologies in European patent law which would involve a similar pre-identification of particular *ordre public* and morality triggers. J Pila, 'Adapting the *ordre public* and morality exclusion of European patent law to accommodate emerging technologies' 38 (2020) Nature Biotechnology 555, 556.

[31] *Transgenic Animals/Harvard*, Second Decision of the Examining Division (1992) OJ EPO 588, Reasons for the Decision [4]; The use of the associative balancing test in *Transgenic Animals/Harvard*, Decision of the Opposition Division (2003) OJ EPO 473 and T0315/03 *Transgenic Animals/Harvard* (2005) OJ EPO 246, (2006) OJ EPO 15 would indicate that such a standard was also applied in the latter decisions (albeit in conjunction with the then new Rule 23d(d) EPC (now Rule 28(1)(d) EPC) exclusion); Warren Jones submits that this is also the standard represented in T0356/93 *Plant Genetic Systems* OJ EPO (1995) 545 Reasons for the Decision [3]–[15], based on the fact that risk-based evidence was taken into account which would not have been relevant if a standard of abhorrence was being applied (A Warren Jones, 'Vital parameters for patent morality – a question of form' (2007) 2 JIPLP 832, 837; see also A Warren Jones, 'Finding a "common morality codex" for biotech – a question of substance' (2008) 39 IIC 638, 652).

ard of universal outrageousness;[32] the abhorrence standard[33] adopted from the Guidelines for Examination in the EPO whereby 'the public in general would regard the invention as so abhorrent that the grant of patent rights would be inconceivable';[34] and the requirement for 'a clear conflict with basic ethical and legal standards'.[35] However, in practice these standards are often used interchangeably and overlappingly.

In the initial decision of the Opposition Division in *Howard Florey/Relaxin*, the terminology of universal outrageousness in conjunction with abhorrence as set out in the Guidelines for Examination in the EPO was utilised.[36] The requirement of a 'clear conflict' with fundamental or basic ethical and legal standards was the chosen terminology in the Opposition Division decisions in *Biocyte*[37] and *Breast & Ovarian Cancer/University of Utah*.[38] However, these

---

[32] *Plant Genetic Systems* EP0242236 Opposition Division (15 February 1993), Reasons for the Decision [3.5] available at https://register.epo.org/espacenet/regviewer under the relevant patent documentation entitled Grounds for the Decision (Annex); *Howard Florey/Relaxin* EP0112149 Opposition Division (1995) OJ EPO 388, Reasons for the Decision [6.2.1].

[33] *Plant Genetic Systems* EP0242236 Opposition Division (15 February 1993), Reasons for the Decision [3.5] available at https://register.epo.org/espacenet/regviewer under the relevant patent documentation entitled Grounds for the Decision (Annex); *Howard Florey/Relaxin* EP0112149 Opposition Division (1995) OJ EPO 388, Reasons for the Decision [6.2.1]; *R v Leland Stanford/ Modified Animal* Opposition Division [2002] 2 EPOR 16, Reasons for the Decision [8]; *Euthanasia Compositions/Michigan State University* EP0516811 Opposition Division (23 May 2000), Reasons for the Decision [2.9] and [2.21].

[34] As reproduced from the earlier versions in Guidelines for Examination in the EPO (Revised Edition, March 2021) Part G – Chapter II 4.1.

[35] *Biocyte* (*Isolation and preservation of fetal and neonatal hematopoietic stem and progenitor cells of the blood*) EP0343217 Opposition Division (21 July 1999), Reasons for the Decision [1.4.3] available in All Documents for EP0343217 as Annex to the Communication at https://register.epo.org/espacenet/regviewer last accessed 28 March 2021; *Breast & Ovarian Cancer/University of Utah* (*Nucleic acid probes comprising a fragment of the 17q-linked breast and ovarian cancer susceptibility gene*) EP0705902 Opposition Division (19 September 2005), Reasons for the Decision [12.2] available in All Documents for EP0705902 as Interlocutory decision in Oral Proceedings at https://register.epo.org/espacenet/regviewer last accessed 28 March 2021.

[36] *Howard Florey/Relxain* Opposition Division (1995) OJ EPO 388, Reasons for the Decision [6.2.1] and [6.3].

[37] *Biocyte* (*Isolation and preservation of fetal and neonatal hematopoietic stem and progenitor cells of the blood*) EP0343217 Opposition Division (21 July 1999), Reasons for the Decision [1.4.3] available in All Documents for EP0343217 as Annex to the Communication at https://register.epo.org/espacenet/regviewer last accessed 28 March 2021.

[38] *Breast & Ovarian Cancer/University of Utah* (*Nucleic acid probes comprising a fragment of the 17q-linked breast and ovarian cancer susceptibility gene*) EP0705902

decisions might also be viewed as conveying a standard of abhorrence, as in both instances the Opposition Division was concerned with considering if there were firm legal or ethical positions to justify the rejection of the patents, which in turn allowed it to quickly dismiss the Article 53(a) EPC submissions.[39]

The standard of unacceptability would appear to require a lower threshold of disapproval to reject an invention, and has been cited as the standard in cases where potential concerns are considered, often in some form of a balancing exercise. The classic example of the standard of unacceptability is the second decision of the Examining Division in *Harvard Oncomouse*, whereby it was considered whether the oncomouse invention would be 'unacceptable' based on the evaluation of relevant concerns in the context of a balancing exercise.[40] However, the Opposition Division in *Leland Stanford* explicitly endorsed the operation of a balancing test but also referred to the requirement for abhorrence in applying Article 53(a) EPC.[41] More recently, abhorrence was the avowed standard in T1553/15 *Rabbit Skin Extract/VanWorld*, but the subsequent, albeit brief, analysis also brought in elements of a balancing exercise.[42] The standard of abhorrence therefore seems to permeate the Article 53(a) EPC jurisprudence, suggesting a subordinate position for unacceptability.

Abhorrence is the standard set out in the Guidelines for Examination in the EPO, which may explain its dominant presence in the decisions of the boards of the EPO, but it is a difficult standard to endorse as it is often manifest in knee-jerk reactions.[43] This is highlighted by Odell-West, who has posited that '[t]o reject the grant of a patent on the basis of an appeal to abhorrence

---

Opposition Division (19 September 2005), Reasons for the Decision [12.2] available in All Documents for EP0705902 as Interlocutory decision in Oral Proceedings at https://register.epo.org/espacenet/regviewer last accessed 28 March 2021.

[39] ibid. Warren Jones also argues that the clear conflict standard amounts to a measure based on abhorrence, despite the absence of a direct reference to abhorrence in the relevant decisions. A Warren Jones, *Taming Scary Monsters with Morality: An Assessment of the Morality Criterion in European and UK Patent Law in the context of Human and Animal Biotechnology* PhD Thesis (Cardiff, 2003) at 210.

[40] The board referred to the invention in this context as opposed to the exploitation of the invention. *Transgenic Animals/Harvard* Second Decision of the Examining Division (1992) OJ EPO 588, Reasons for the Decision [4].

[41] *R v Leland Stanford/ Modified Animal* [2002] 2 EPOR 16, Reasons for the Decision [8].

[42] This test was applied here in respect of animals but it derives from decisions dealing with genetic and cellularly modified animals: (28 September 2020), Reasons for the Decision [2.4]–[2.5]. The facts concerned the use of rabbits to make pharmaceutical substances. The rabbits were injected with a virus to inflame their skin, causing suffering before the rabbit was killed and the skin 'harvested'.

[43] Guidelines for Examination in the EPO (Revised Edition, March 2021) Part G – Chapter II 4.1.

or reprehensibility without moral reasoning is an appeal to bare prejudice'.[44] Beyleveld and Brownsword similarly contend that the 'abhorrence' standard is problematic as it 'evokes a fairly strident chorus of disapproval' and therefore constitutes a 'cultural morality', as opposed to their preferred standard of 'critical cultural morality'.[45] They propose the application of a 'critical cultural morality' which requires more than just an instinctive reaction to a technology and involves '"moral" judgments' to satisfy 'tests of sincerity and rationality'.[46] Further, they consider the emphasis on the prevalence or 'quantity' of a moral consensus unhelpful, as if something is 'intrinsically immoral' then 'the presence or absence of an overwhelming consensus condemning such patents is irrelevant'.[47]

The problem highlighted by Beyleveld and Brownsword is perhaps exacerbated by the emphasis on a *European* consensus in the main morality definition provided by the boards of the EPO and suggested by Recital 39 of the Biotech Directive. This in turn has led to discussions as to how to engage with individual state perspectives in the context of Article 53(a) EPC, which could potentially paralyse meaningful engagement with the exclusion by requiring, to varying degrees, the identification of specific viewpoints in respect of any contentious inventions in each Contracting State of the EPC and deciding if some level of majority or qualified majority could justify the exclusion of the patent application.[48] It is difficult to sustain any such approach in the context of a body making patent law decisions that encompass numerous domestic jurisdictions, many of which will not have reached a consensus on any *ordre public*

---

44 A Odell-West, 'Invention and the human embryo' (2020) 1 IPQ 9.

45 D Beyleveld and R Brownsword, 'Patenting human genes: legality, morality, and human rights' in JW Harris (ed), *Property Problems: From Genes to Pension Funds* (Kluwer Law International, 1997) 9, 21.

46 D Beyleveld and R Brownsword, 'Patenting human genes: legality, morality, and human rights' in JW Harris (ed), *Property Problems: From Genes to Pension Funds* (Kluwer Law International, 1997) 9, 21.

47 D Beyleveld and R Brownsword, 'Patenting human genes: legality, morality, and human rights' in JW Harris (ed), *Property Problems: From Genes to Pension Funds* (Kluwer Law International, 1997) 9, 22.

48 P Torremans, 'A transnational institution confronted with a single jurisdiction model: guidance for the EPO's Implementation of the Directive from a private international law perspective' in A Plomer and P Torremans (eds), *Embryonic Stem Cell Patents: European Law and Ethics* (Oxford University Press, 2009) 271, 298. See also J Straus, 'Patenting of human genes and living organisms – the legal situation in Europe' in F Vogel and R Grunwald (eds), *Patenting of Human Genes and Living Organisms* (Springer, 1994) 12, 26; A Warren Jones, 'Finding a "common morality codex" for biotech – a question of substance' (2008) 6 (39) IIC 638, 652; A Warren Jones, 'Vital parameters for patent morality – a question of form' (2007) 2 (12) JIPLP 832, 844:

and/or morality issues pertaining to new technologies. Further, it would also seem to be limited by the proviso of the exclusion which specifically limits resort to national legislation for definitive answers.[49] Necessarily, this suggests that the provision requires individual analysis of the relevant issues by a relevant authority (ideally with the assistance of a specialised ethics committee) in order to consider the *ordre public* and/or morality issues directly. In this regard the standard of unacceptability is preferable in determining whether to deny patent protection under Article 53(a) EPC, as it seems to provide for a more qualitative approach by the decision makers, even where nuanced concerns are involved.

3.2 Methodologies

The methods used in conjunction with the identified standards cause further confusion in the interpretation of Article 53(a) EPC. Two main assessment methods adopted by the boards of the EPO have been identified by Warren Jones: the utilitarian/consequentialist test and the rebuttable presumption test.[50] The utilitarian approach has been presented as the principal approach in cases concerning animals,[51] and may also apply in plant-related cases.[52] However, in the second decision of the Technical Board of Appeal in T0315/03 *Harvard Oncomouse* this initially straightforward approach was undermined by the use of two balancing tests (the original test formulated in T19/90 *Harvard Oncomouse* in conjunction with the new Rule 28(1)(d) (then Rule 23d(d) test) as well as an overall assessment of any additional relevant issues).[53]

The first test weighs up environmental risks and animal suffering against likely usefulness to humankind, whereas the latter test omits environmental

---

[49] 'such exploitation shall not be deemed to be so contrary merely because it is prohibited by law or regulation in some or all of the Contracting States.'

[50] A Warren Jones, 'Finding a "common morality codex" for biotech – a question of substance' (2008) 6 (39) IIC 638, 653 and A Warren Jones, 'Vital parameters for patent morality – a question of form' (2007) 2 (12) JIPLP 832, 834.

[51] See for example T0315/03 *Transgenic Animals/Harvard* (2005) OJ EPC 246/ (2006) OJ EPO 15, Reasons for the Decision [10]; T19/90 *Transgenic Animals/ Harvard* (first decision of the Technical Board of Appeal) [1990] EPOR 501, Reasons for the Decision [5] and *Gene Trap/Artemis* T0606/03 (12 January 2005), Reasons for the Decision [19] available at www.epo.org/law-practice/case-law-appeals/recent/ t030606eu1.html last accessed 28 March 2021.

[52] See T0356/93 *Plant Genetic Systems* OJ EPO (1995) 545, Reasons for the Decision [18.8] where the use of a balancing test is rejected in this instance due to the lack of sufficient environmental risks evidence.

[53] T0315/03 *Transgenic Animals/Harvard* (2005) OJ EPO 246/ (2006) OJ EPO 15, Reasons for the Decision [10] and [13.2].

concerns and requires a likely 'substantial medical benefit' for humans.[54] The tests are both too similar, yet perceptibly distinct, to do anything but unnecessarily complicate a weighing up of relevant concerns. This approach was further compounded by the inclusion of an additional general analysis under Article 53(a) EPC in T0315/03 *Harvard Oncomouse*,[55] entailing a three-pronged assessment, which was subsequently adopted in T0606/03 *Gene Trap/Artemis*.[56]

Prima facie, a utilitarian test seems attractive, involving as it does a weighing up of all the relevant pros and cons to determine if the commercial exploitation of the invention would contravene *ordre public* and/or morality in Europe. However, when used in conjunction with other tests, as in T0315/03 *Harvard Oncomouse* and T0606/03 *Gene Trap/Artemis*, the practical utility of the test is undermined. This position is undoubtedly complicated by the need to indirectly maintain the status quo with EU member states based on the Biotech Directive, thereby limiting the EPO's power to formulate a single balancing test in the context of genetically modified animals. Notwithstanding this, the EPO's typical dismissive approach to opposing arguments and evidence undermines the functionality of balancing exercises in practice. This is apparent in the treatment of risk-based evidence regarding potential environmental damage in T0356/93 *Plant Genetic Systems*. The board found that the fact that the evidence referred to 'the possibility' of 'some undesired, destructive events' undermined the appellant's arguments.[57] This was in spite of the fact that the respondents were willing to concede that at least some of the specified risks might 'occur to some extent'.[58] Ultimately, a lack of 'sufficient evidence of actual disadvantages' led the board to conclude that no balancing exercise was possible.[59] On the other hand, evidence presented by the patent applicant of substantial medical benefit in T0315/03 *Harvard Oncomouse* was readily accepted, and indeed was even found to 'at the very least be inferred from

---

[54] Rule 28(1)(d) excludes 'processes for modifying the genetic identity of animals which are likely to cause them suffering without any substantial medical benefit to man or animal, and also animals resulting from such processes'. This is identical to the originating provision in Article 6(2)(d) of the Biotech Directive.

[55] T0315/03 *Transgenic Animals/Harvard* (2005) OJ EPO 246/ (2006) OJ EPO 15, Reasons for the Decision [13.2.10]–[13.2.21].

[56] T0606/03 *Gene Trap/Artemis* (12 January 2005), Reasons for the Decision [12]–[15] 12 January 2005) available at www.epo.org/law-practice/case-law-appeals/recent/t030606eu1.html_last accessed 28 March 2021.

[57] T0356/93 *Plant Genetic Systems* (1995) OJ EPO 545, Reasons for the Decision [18.6].

[58] ibid.

[59] ibid [18.8].

the patent itself'.[60] This was notwithstanding the fact that this analysis also entailed a forward-looking assessment of an uncertain future outcome.

A rebuttable presumption approach involves a presumption that the invention is patentable and not excluded by Article 53(a) EPC unless evidence is presented which establishes a contrary position. In contrast to balancing the pros and cons, this essentially involves considering any concern(s) – dependent on what the board considers relevant and/or based on what has been submitted on opposition – and examining whether any one of them would result in the commercial exploitation of the invention being abhorrent or unacceptable.[61] The rebuttable presumption approach is epitomized in cases such as *Howard Florey/Relaxin*, where submitted concerns regarding the patentability of the isolated hormone relaxin were dispatched as being not sufficiently important or incorrectly conceived in order to allow for patentability of the isolated human material.[62] This led to arguments in respect of the potential for slavery due to the commodification of human material being rejected as representative of a 'fundamental misunderstanding of the effects of a patent'.[63] On the other hand, in T0149/11 *Slaughterhouse* a similar approach was used to prevent the patenting of an invention where a poorly drafted claim could be perceived as including slavery within its ambit. In the latter case, the possibility of slavery as part of the invention was so abhorrent as to merit a decision of unpatentability.[64]

The application of the rebuttable presumption test is therefore very much dependent on the existence of a widely accepted, prescribed rejection of a key aspect of the commercial exploitation of the invention. Further, the contrasting approach to arguments in respect of slavery in *Howard Florey/Relaxin* and *Slaughterhouse* indicates a vast scope for varying approaches to the test. But it is particularly problematic in the arena of new developing technologies

---

[60] T0315/03 *Transgenic Animals/Harvard* (2005) OJ EPO 246/ (2006) OJ EPO 15, Reasons for the Decision [13.2.2].

[61] A Warren Jones, 'Finding a "common morality codex" for biotech – a question of substance' (2008) 6 (39) IIC 638, 653 (Warren Jones argues that the rebuttable presumption is intrinsically tied to the standard of abhorrence). See also A Warren Jones, 'Vital parameters for patent morality – a question of form' (2007) 2 (12) JIPLP 832, 834.

[62] *Howard Florey/Relaxin* Opposition Division (1995) OJ EPO 388, Reasons for the Decision [6.1].

[63] *Howard Florey/Relaxin* Opposition Division (1995) OJ EPO 388, Reasons for the Decision [6.3.3].

[64] T0149/11 *Method and device for processing a slaughtered animal or part thereof in a slaughterhouse* EP819381 (24 January 2013) Reasons for the Decision [2.5] – [2.6] available at www.epo.org/law-practice/case-law-appeals/recent/t110149eu1.html last accessed 28 March 2021.

which society and the law have yet to fully evaluate, let alone reach definitive decisions of abhorrence or indeed unacceptability on their commercial exploitation. The application of a utilitarian test in relation to human material seems to have been outrightly rejected by the Technical Board of Appeal in T1374/04 *WARF* as distasteful[65] but appears to present a more realistic, practical all-round framework for evaluating evidence than the more favoured rebuttable presumption methodology. However, the utilitarian approach would only function successfully if a more serious effort was applied to the evaluation of evidence under Article 53(a) EPC.

4. COMMERCIAL EXPLOITATION

In addition to the lack of clarity regarding the standards and methodologies utilised by the boards of the EPO, there is lingering uncertainty regarding the scope of the exclusion. Therefore, despite the present wording of the provision referring to 'commercial exploitation', the EPO refuses to consider the economic impact of patenting. This is manifest in the decision of G-0001/98 *Transgenic Plant/Novartis II*, where the EBA set out that the EPO had 'not been vested with the task of taking into account the economic effects of the grant of patents in specific areas and of restricting the field of patentable subject-matter accordingly'.[66]

Similarly, in the *Breast & Ovarian Cancer/University of Utah* decisions, objections regarding the inhibiting effect of patents on research and costs for patients were dispelled *en masse* as incorrectly directed on the effect of the patents.[67] Notwithstanding this, Moufang suggests that there might be rare situations where 'the economic exclusionary position afforded by the patent' might need to be considered.[68] There have also been recent calls for a broad-

[65] T1374/04 *Wisconsin Alumni Research Foundation (WARF)* OJ EPO (2007) 313 Reasons for the Decision [55].

[66] G_0001/98 *Transgenic Plant/Novartis II* (2000) OJ EPO 111, Reasons for the Decision [3.9].

[67] T1213/05 *Breast & Ovarian Cancer/University of Utah* (27 September 2007) Reasons for the Decision [53] available at www.epo.org/law-practice/case-law-appeals/recent/t051213eu1.html last accessed 28 March 2021; *Breast & Ovarian Cancer/University of Utah* (*Nucleic acid probes comprising a fragment of the 17q-linked breast and ovarian cancer susceptibility gene*) EP0705902 Opposition Division (19 September 2005), Reasons for the Decision [12.2] available in All Documents for EP0705902 as Interlocutory decision in Oral Proceedings at https://register.epo.org/espacenet/regviewer last accessed 28 March 2021.

[68] R Moufang, 'Patenting of human genes, cells and parts of the body? The ethical dimensions of patent law' (1994) 25 (4) IIC 487, 504.

ening of the scope of Article 53(a) EPC to include the economic and access effects of patents.[69]

Traditionally, other concerns have also been found to fall outside the remit of Article 53(a) EPC. Indeed, even though the old wording of Article 53(a) EPC expressly set the domain of the test as 'the publication or exploitation' of the invention, publication does not seem ever to have been contemplated as relevant to the exclusion. In T0866/01 *Euthanasia Compositions* the Technical Board of Appeal stated that it was 'manifestly absurd to suggest that the routine act of publication could retrospectively become the reason for refusing a patent application or revoking a granted patent'.[70]

There appears to be an equally reticent approach in considering developmental issues in respect of an invention under Article 53(a) EPC *generally*. In T1213/05 *Breast & Ovarian Cancer/University of Utah* the board refused to take into account issues of consent to the obtaining of human biological material.[71] This confirmed the earlier decision of the then European Court of Justice in C-377/98 *Kingdom of the Netherlands v European Parliament and Council of the European Union*, whereby it was set out that notwithstanding the existence of Recital 26, issues such as consent were 'clearly misplaced as against a directive which concerns only the grant of patents and whose scope does not therefore extend to activities before and after that grant, whether they involve research or the use of the patented products'.[72] Similarly, in T0866/01 *Euthanasia Compositions/Michigan State University* the Technical Board of Appeal expressly stated that the use of experiments in the making of the invention was not relevant to the Article 53(a) EPC test.[73]

69 See for example J Pila, 'Adapting the *ordre public* and morality exclusion of European patent law to accommodate emerging technologies' 38 (2020) Nature Biotechnology 555, 556 and K Walsh and N Hawkins, 'Expanding the role of morality and public policy in European patent law' in P Torremans (ed), *Intellectual Property Law and Human Rights* (4th edition, Wolters Kluwer, 2020) 19 (online) available at http://hdl.handle.net/10871/40561 last accessed 28 March 2021.

70 T0866/01 *Euthanasia Compositions/Michigan State University* EP (11 May 2005), Reasons for the Decision, [6.15] available at www.epo.org/law-practice/case-law-appeals/recent/t010866eu1.html last accessed 28 March 2021.

71 T1213/05 *Breast & Ovarian Cancer/University of Utah* (27 September 2007), Reasons for the Decision [48]–[51] available at www.epo.org/law-practice/case-law-appeals/recent/t051213eu1.html last accessed 28 March 2021.

72 [2001] ECR I-7079 [79]. Recital 26 sets out: '[w]hereas if an invention is based on biological material of human origin or if it uses such material, where a patent application is filed, the person from whose body the material is taken must have had an opportunity of expressing free and informed consent thereto, in accordance with national law.'

73 T0866/01 *Euthanasia Compositions/Michigan State University* EP (11 May 2005), Reasons for the Decision, [6.8 (b)] available at www.epo.org/law-practice/case-law-appeals/recent/t010866eu1.html last accessed 28 March 2021.

In spite of this, academic commentators have argued that ignoring past developmental transgressions is untenable under Article 53(a) EPC.[74] Moufang, who has served on the boards of appeal on numerous occasions, suggested that developmental aspects of the invention could come into play if 'decisively characterised by an ethically dubious procedure':[75]

> If an invention gains a decisive lead in time, e.g. during the race for priority, by undertaking experiments that are forbidden because of their dangerousness or their otherwise unacceptable ethical implications, then adequate sanctions must be provided under patent law. If a patent were to be granted to persons who ignore fundamental ethical principles, the interests of all other participants in the research and development process would be flagrantly neglected.[76]

A limited acceptance of developmental concerns seemed to receive support in the application of the express exception relating to 'uses of human embryos for industrial or commercial purposes' under Article 6(2)(c) of the Biotech Directive/Rule 28(1)(c) EPC. In this regard, the Court of Justice of the EU (CJEU) set out that uses of human embryos (which were not beneficial to it: for example, destruction) in the development of an invention should be excluded from patent protection in C-34/10 *Brüstle v Greenpeace eV*.[77] This was perhaps best articulated by Advocate General Bot, who set out that it would not be possible to patent an invention which was created using organs that had been obtained through criminal activity which contravened human dignity.[78] A concern for developmental issues was manifest in the decision of the CJEU to set out that even historic destruction of the human embryo was excluded under the Directive. Therefore, '[t]he fact that destruction may occur at a stage long before the implementation of the invention' was deemed to be 'irrelevant'.[79]

---

[74] M Rowlandson, 'The ordre public and morality exception and its impact on the patentability of human embryonic stem cells' (2010) EIPR 60, 75; D Beyleveld and R Brownsword, *Mice, Morality and Patents* (Common Law Institute of Intellectual Property, London, 1993) 51; S Sterck and J Cockbain, 'Assessing the morality of the commercial exploitation of inventions concerning uses of human embryos and the relevance of moral complicity: comments on the EPO's WARF Decision' (2010) 7 (1) SCRIPTed 83, 100–1.

[75] R Moufang, 'Patenting of human genes, cells and parts of the body? The ethical dimensions of patent law' (1994) 25 (4) IIC 487, 504.

[76] R Moufang, 'Patenting of human genes, cells and parts of the body? The ethical dimensions of patent law' (1994) 25 (4) IIC 487, 504.

[77] C-34/10 *Brüstle v Greenpeace eV* ECLI:EU:C:2011:669 [52].

[78] C-34/10 *Brüstle v Greenpeace eV* ECLI:EU:C:2011:669, Opinion of the Advocate General [106].

[79] C-34/10 *Brüstle v Greenpeace eV* ECLI:EU:C:2011:669 [49].

The approach in *Brüstle* was subsequently adopted in the boards of the EPO in respect of Rule 28(1)(c), even though this required a change of policy towards such inventions. Prior to the *Brüstle* decision, the EPO had maintained a practice of allowing inventions dependent on historic destruction of human embryos to be patentable if there was a human embryonic stem cell line available as a pre-existing source of stem cells, thereby not requiring new destruction of human embryos in order to make/exploit the invention.[80] This approach was altered again in response to the decision in C-364/13 *International Stem Cell Corporation v Comptroller General of Patent, Designs and Trade Marks (ISCC)*, where it was deemed that parthenotes did not come within the EU definition of human embryo.[81] This facilitated the grant of patents for inventions where human embryonic stem cells were required but stem cell lines created by the destruction of parthenotes (rather than human embryos) were available at the priority or filing date.[82]

The decision in *Brüstle* and the approach the boards of the EPO subsequently adopted in response may suggest a refocusing of the *ordre public* and morality exclusion, but it is unclear if this applies to Rule 28(1)(c) specifically, the new rules generally or the provision overall. Further, it is possible that the focus on developmental concerns is only related to the 'commercial exploitation' of the invention and that broader concerns not related to the reproducibility of the invention are still not relevant. In the subsequent decision of *Slaughterhouse*, for instance, the Technical Board of Appeal emphasised that the focus of the test was not the invention itself but its 'exploitation', although it did not expand on what this meant specifically.[83] In this regard the precise limits of Article 53(a) EPC remain elusive, which in turn frustrates the patent applicant, inventor and opponent.

80 The date seemed to be fixed at May 2003. See P Webber, '*Brüstle*: what will happen next?' (29 October 2011) https://ipkitten.blogspot.com/2011/10/brustle-what-will-happen-next.html last accessed 28 March 2021.

81 C-364/13 *International Stem Cell Corporation v Comptroller General of Patent, Designs and Trade Marks* ECLI:EU:C:2014:2451 [38].

82 This date has been set at 5 June 2003. T0385/14 *In vitro differentiated cardiomyocytes/Axiogenesis* (11 September 2019), Reasons for the Decision [3] available at www.epo.org/law-practice/case-law-appeals/recent/t140385eu1.html last accessed 28 March 2021.

83 T0149/11 *Method and device for processing a slaughtered animal or part thereof in a slaughterhouse* EP819381 (24 January 2013) Reasons for the Decision [2.1] available at www.epo.org/law-practice/case-law-appeals/recent/t110149eu1.html last accessed 28 March 2021).

5. EXPRESS EXCLUSIONS: ARTICLE 6(2)/RULE 28

In comparison to the traditional reticent approaches of the boards of the EPO, the combined EPO and CJEU approaches to inventions involving human embryonic stem cells initially appear to represent a more practical engagement with *ordre public* and morality. However, the structure and interpretation of the express exclusions of Article 6(2)(c) of the Biotech Directive adopted as Rule 28(1) EPC present additional problems for the application of Article 53(a) EPC.

As set out above, the introduction under Rule 28(1)(d) of a specific (albeit not identical) exclusion relating to transgenic animals designed to suffer added an extra layer of complexity to the existing approach to animal suffering in T0315/03 *Harvard Oncomouse*. Rule 28(1)(c) is in theory less problematic to apply, requiring as it does that the board merely decide if the invention contravenes a specific exclusion. However, it has been subject to varying interpretations in the CJEU and the EPO. This is in part due to the changing interpretation of human embryo (to exclude parthenotes) in *ISCC* and the potential confusion regarding the scope of the exclusion. However, there are additional problems with the uncertain justification of the exclusion, which might suggest that an individual assessment weighing up the pros and cons of the uses of human embryos might be a more ethically valid exercise than the existence of a firm obstacle supposedly founded on human dignity.

In applying Rule 28(1)(c) there is no scope to consider broader policy discussions.[84] Therefore, as Odell West notes, in the redrawing of the definition of human embryo to exclude parthenotes, the question of parthenote suffering which is likely to be similar to the suffering of human embryos is not considered at all.[85] In the 2019 decision of T0385/14 *In vitro differentiated cardiomyocytes/Axiogenesis*, where the Technical Board of Appeal agreed that the commercial exploitation of the relevant invention no longer fell within the scope of Rule 28(1)(c) due to the changed definition of human embryo to exclude parthenotes, no discussion ensued as to whether there were additional Article 53(a) EPC issues (such as parthenote suffering) to take into account.[86] The use of an absolute exclusion in Rule 28(c) therefore seems to detract from

84 K Walsh and N Hawkins, 'Expanding the role of morality and public policy in European patent law' in P Torremans *Intellectual Property Law and Human Rights* (4th edition, Wolters Kluwer, 2020) 14 (online) available at http://hdl.handle.net/10871/40561 last accessed 28 March 2021.

85 A Odell-West, 'Invention and the human embryo' (2020)1 IPQ 1, 10–11.

86 *In vitro differentiated cardiomyocytes/Axiogenesis* (11 September 2019) Reasons for the Decision [4]-[5] available at www.epo.org/law-practice/case-law-appeals/recent/t140385eu1.html last accessed 28 March 2021.

a more complete analysis of *ordre public* and morality concerns. This is in contrast to the, albeit flawed, analysis of broader *ordre public* and morality concerns in situations where Rule 28(d) has been engaged.

The exclusion relating to human embryos has been criticised on a fundamental level due to the lack of certainty regarding the status of the human embryo in the development of human life and its subsequent removal of broader considerations in respect of the potential benefits of inventions involving the use of human embryos.[87] There may also be issues with the underlying justification for 'processes for modifying the germ line genetic identify of human beings' under Article 6(2)(c)/Rule 28(1)(b) in respect of the potential for such processes in the eradication of genetic disease or indeed improving human life generally. The presence and application of the express exclusions therefore not only fail to clarify the interpretation of *ordre public* and morality in the EPO but potentially impede the application of principles of *ordre public* and morality.

6. CONCLUSION

It is apparent that the approach to *ordre public* and morality in the boards of the EPO, in conjunction with the additional rules introduced in response to the Biotech Directive, have resulted in a complicated, unsuccessful approach to Article 53(a) EPC. It is this convoluted approach to *ordre public* and morality that presents an obstacle to the patentability of emerging technologies, as opposed to the key provision itself.

However, the boards of the EPO need to fully engage with *ordre public* and morality. First, this is expedient in order to develop a consistent approach to the issue. It is apparent from the analysis here that the present approach is inconsistent and encompasses a proliferation of tests and standards. This is problematic for all participants in the patenting process. Opponents cannot be sure what evidence to present to put forward their case; inventors and applicants cannot be certain that they will not reach an obstacle in the EPO based on *ordre public* and morality even if they comply with research regulations. Neither can they be certain as to whether, having invested in developing and patenting the invention, its exploitation might be limited further downstream after the patenting process.[88]

An isolationist approach to patenting, whereby the process is seen as other to broader regulation of research, development and exploitation as opposed

---

[87] A Odell-West, 'Invention and the human embryo' (2020)1 IPQ 1, 6.

[88] A McMahon, 'Biotechnology, health and patents as private governance tools: the good, the bad and the potential for ugly?' 3 (2020) IPQ 161.

to part of the research, development and exploitation process, undermines the justification of patent rights as incentives to inventiveness. By engaging appropriately with *ordre public* and morality the patent office could place itself at the centre of new technology, as well as lead the way in not only incentivising inventions but incentivising *ethical* inventions through how the exclusion is operated. The scientific background of patent examiners could benefit a debate that risks becoming mired in confusion in non-scientific contexts. More expansive engagement with Article 53(a) EPC would be best served by the incorporation of an ethics committee at first instance, but could also encompass ongoing public consultation and evaluation.[89] In this regard, the EPO could set out guidelines regarding emerging and new technologies, thus identifying issues that may arise before applications arrive in the boards of the EPO. The EPO could thus lead the debate in respect of new technologies as opposed to playing catch-up, as it found itself doing in the human embryonic stem cell cases. If successfully achieved, such an approach could indeed lead to Article 53(a) EPC only requiring application in 'rare and extreme' cases.[90]

---

[89] Alternative mechanisms to deal with ordre public and morality have been proposed by M O'Sullivan, *Biotechnology, Patents and Morality: A Deliberative and Participatory Paradigm for Reform* (Routledge, 2020) and J Pila, 'Adapting the *ordre public* and morality exclusion of European patent law to accommodate emerging technologies' 38 (2020) Nature Biotechnology 555.

[90] Guidelines for Examination in the EPO (Revised Edition, March 2021) Part G – Chapter II 4.1.

6. Invoking the public interest in the European patent system

Karen Walsh[1]

1. INTRODUCTION

The European patent system is based on a balance between private rights and access. This balance is apparently achieved through the patent bargain, where private rights are only granted if the information revealed in a patent application provides a corresponding benefit to society. It is said that the short-term exclusivity granted to patentees boosts innovation and benefits the public in the long term. This balance between private rights and access has been considered in European patent legislation through substantive patent law provisions and procedural requirements.

Both the encouragement of innovation and having access to that information are in the public interest, but in practice the balance often shifts in favour of the rightsholder and away from the public. In this chapter it is argued that the public interest in access has been overwhelmed by practices that shift the balance towards the granting of patents, and thus towards patentees. This has occurred through a reliance on the public interest being baked into the European patent system, coupled with narrow interpretations of exceptions and exclusions to patentability.

This chapter will first examine the underlying patent balance and the public interest. This will be followed by an uncovering of the public interest within the legislation itself and in its current interpretation by numerous institutions. It is posited that due to the inherent nature of its consideration within legislation, a more detailed analysis of the public interest is missing from interpretation. If it is assumed that the public interest has been considered, this may restrict its

[1] The author would like to thank the editor, Naomi Hawkins, and Mathilde Pavis for comments on an earlier draft of this chapter. A version of this chapter was presented at ATRIP, Nashville 2019 and at the Patents and Emerging Technologies Workshop, London 2019. The author is grateful for the feedback received at these events. All links last accessed on 3 August 2021.

further inclusion in deliberations. As a result, narrow interpretations are implemented without further consideration of the public interest. This argument is subsequently analysed by examining the relevant and timely interpretation of the exception to patentability relating to essentially biological processes by the Boards of Appeal of the European Patent Office. It is concluded that an open and clear avenue is required in order to invoke the public interest whenever necessary.

Invoking the public interest is particularly key for emerging technologies. What should be protected in the first place, and to what extent, should be questioned, with the public interest as the focal point in decision-making. There should be an opportunity to question the balance at any stage of the patent life cycle to ensure that the public interest in access is not overshadowed.

It should be noted that 'the public interest' is an elusive term. In its general conception, it fails to recognise and account for the reality of numerous publics with numerous interests.[2] In this chapter, viewed from the lens of the patent balance, it is argued that the manner in which the patent balance is put into practice is failing publics, particularly small and medium sized enterprises and individuals working in particular fields, and society in general to various degrees. This chapter mostly makes reference to European patent law and case law. However, as European legislation has been implemented in many national laws, and patent systems in many jurisdictions rely on the same balance, the points made herein are widely applicable.

2. THEORETICAL UNDERPINNINGS OF THE PATENT BALANCE AND THE PUBLIC INTEREST

The balance constructed and ostensibly preserved by the patent system is conceptually straightforward. Although in reality the situation is complex, broken down into simple terms this balance exists to maintain an equilibrium between private rights on one side and access on the other. The public interest emerges on both sides. This is seen in operation through the patent bargain.

For the successful patentee, the system rewards them with an exclusive right over their invention. This exclusive right prevents any third party from making, using, offering for sale or keeping that invention in the territory for which the right is granted.[3] It restricts access. Such protection is granted only

---

[2] Karen Walsh, Andrea Wallace, Mathilde Pavis, Natalie Olszowy, James Griffin and Naomi Hawkins, 'Intellectual Property Rights and Access in Crisis' (2021) 52(4) IIC 379.

[3] These are the minimum standards of protection for all members of the World Trade Organization (WTO) as outlined by the Agreement on Trade Related Aspects of

if the invention meets the requirements of patentability.[4] It must be worthy of patent protection, or in other words, worthy of restricting access.

Patent protection includes both costs and benefits. Innovation generally results in advancements in technologies that will be valued by society. New inventions provide the public with numerous advantages, such as information, new technologies and medical advancements, to name a few. However, once an invention is patent protected, access is restricted, and it cannot be used unless permission from the patentee has been received, the patent lapses, it is declared invalid or the term of protection has come to an end. Therein lies the bargain and balance – an invention is granted patent protection only if it leads to advancements for society, but the right holder can restrict access to that invention for a limited time.

The justifications for the patent system have been discussed at length elsewhere.[5] Briefly, patents are said to incentivise innovation. It is argued that without the opportunity of a reward in the form of a patent right, there would be no incentive to innovate and no opportunity to recoup investment.[6] It is also argued that without the patent system, information would be kept secret and as a result society would benefit less from advancements in technology – that by having a patent system, this information is disclosed and available for use once patent protection ends.[7] Following these arguments, this time-limited restriction is seen as necessary to bring a benefit to society in the short term by disclosing and providing new information, as well as in the long term once the patent has expired. Here we can see the public interest operating on both sides of the bargain.

However, whether the patent system is required to encourage innovation is often questioned.[8] It has been argued that there are many other schemes that would work just as well if not better than the patent system, and in a more equi-

---

Intellectual Property Rights 1994 (TRIPS) under Article 28. 164 countries are currently members of the WTO.

4 Article 27 TRIPS sets out the minimum requirements for patentable subject-matter (novelty, inventive step and industrial application) and provides indications on what may be excluded from patent protection.

5 For one, see: Robert Merges, *Justifying Intellectual Property* (Harvard University Press 2011).

6 Dominique Guellec, 'Patents as an Incentive to Innovate' in Dominique Guellec and Bruno van Pottelsberghe de la Potterie (eds), *The Economics of the European Patent System: IP Policy for Innovation and Competition* (OUP 2007) 49.

7 Guellec, 49.

8 These debates have been summarised by Rochelle Dreyfuss, 'In Praise of an Incentive-Based Theory' in Rochelle Cooper Dreyfuss and Elizabeth Siew-Juan Ng (eds) *Framing Intellectual Property Law in the 21st Century: Integrating Incentives, Trade, Development, Culture, and Human Rights* (CUP 2018) 3–11.

table manner.[9] Whether the system achieves the purported balance or bargain that justifies it has also been questioned.[10] That it does not is particularly apparent when seen alongside considerations of the existing development divide and wealth gap.[11] It is difficult to see how the public interest in access is being balanced against the scope and type of private rights that are being granted.

It is unlikely that the patent system will be replaced in the near future, if at all, and therefore it is important that the current system operates as effectively and equitably as possible. A true balance between private rights and access, in a way that elevates the public interest, is crucial to delivering this goal. As stated by Rebecca Tushnet, 'the concept of public interest in intellectual property theory generally seems to mean a thumb on the scales against private control in certain arguments about good policy, as well as concern for distribution and not just for maximising utility'.[12] For example, it can be used to push back against over-patenting when subject-matter is being too broadly construed. It can be used to argue that an invention should not be patentable at all. However, it is argued by this author that the public interest in accessing innovations could and should be considered more extensively.

When considering emerging technologies, this is of particular importance. Broadly speaking, progress in science and technology is rapidly advancing and showing no signs of deceleration. Both the encouragement of innovation and having access to that information are in the public interest, but what should be protected in the first place, and to what extent, should be questioned, with the public interest as the focal point in decision-making at each stage in the patent life cycle.

In order to consider where and how the public interest could be invoked more readily within the patent system for both existing and emerging technol-

---

[9] See, for example: Dean Baker, Arjun Jayadev and Joseph Stiglitz, 'Intellectual Property and Development: A better set of approaches for the 21st Century' (2017) available at: http://ip-unit.org/wp-content/uploads/2017/07/IP-for-21st-Century-EN.pdf; Brian Wright, 'The Economics of Invention Incentives: Patents, Prizes, and Research Contracts' (1983) 73(4) The American Economic Review 691; James Love and Tim Hubbard, 'Prizes for Innovation of New Medicines and Vaccines' (2009) 18 Annals Health L 155.

[10] David Vaver, 'Intellectual Property: Still a "Bargain"?' (2012) 34(9) EIPR 579; Mario Biagioli, 'Weighing Intellectual Property: Can We Balance the Social Costs and Benefits of Patenting?' (2019) 57(1) History of Science 140.

[11] Margaret Chon, 'Intellectual Property and the Development Divide' (2006) 27 Cardozo Law Review 2821; Peter Drahos, 'Reclaiming the Patent Social Contract' in *The Global Governance of Knowledge* (CUP 2010) 27–32.

[12] Rebecca Tushnet, 'Intellectual Property as a Public Interest Mechanism' in Rochelle Cooper Dreyfuss and Justine Pila (eds), *The Oxford Handbook of Intellectual Property Law* (Oxford University Press 2017).

ogies, this chapter will next outline where the balance currently emerges in the current system and how the public interest is incorporated into legislation and decision-making.

3. CONSIDERATIONS OF THE PATENT BALANCE AND THE PUBLIC INTEREST IN LEGISLATION AND DECISION-MAKING

This chapter will focus on two key points at which the balance between private rights and access is considered in the European patent system. First is in the criteria for patentability contained in legislation. Second is in the interpretation of those criteria. These points are inextricably linked and play a prominent role in how the current system invokes the public interest. How these criteria are interpreted is often based on why the provision was first introduced. Further, this interpretation has an impact on how the system operates in practice, which subsequently has an impact on the public.

3.1 The Legislation

The balance that patent law seeks to achieve is manifest in both the substantive criteria for patentability and the procedural patent application requirements. It is also apparent in areas such as infringement and licensing, which have been elaborated on in other chapters of this book. This chapter will focus on the criteria for patentability. The substantive provisions include patentable subject matter, novelty, inventive step, industrial application and the exceptions to patentability. The procedural requirements include sufficiency, clarity, unity and added matter. Each will be briefly taken in turn to determine the extent to which the public interest is built into the patent system.

3.1.1 Patentable subject-matter

The subject-matter of a patent application must be an invention that is not excluded from patentability. Subject-matter will be excluded if it is: a discovery, scientific theory or mathematical method; a literary, dramatic, musical or artistic work; an aesthetic creation; a scheme, rule or method for performing mental acts, playing games or doing business; a computer program; or the presentation of information, as such.[13]

Excluding such subject-matter from patent protection has a clear public interest rationale. For some, protection would be unjust as the subject-matter should be free for all to use. For others, the subject-matter is protected else-

[13] Article 52 EPC.

where. For all, the subject-matter does not meet the requirement of being an invention. The main requirement for subject-matter to be considered an invention is that it is technical and not excluded. Thus, the common theme behind these exclusions is that they are mostly abstract rather than specific and technical. This requirement ensures that a patent is not granted for subject-matter without a technical character or contribution, depending on the jurisdiction.[14] This should ensure that society is only restricted from accessing the invention if it contributes something of a technical nature or advancement.

3.1.2 Novelty

The invention must also be new.[15] It cannot form part of the state of the art, or in other words, it cannot have been made available to the public before the priority date of the application. The state of the art is incredibly broad. It includes disclosures that are either written or oral, by use, or any other manner.[16] There are some restrictions on the state of the art. For example, disclosures made in a breach of confidence are not included, mosaicking pieces of prior art together is not permitted and the disclosure must be one that enables the person skilled in the art to make the invention work. However, the rationale behind the provision clearly has the public interest in mind. The novelty requirement operates to ensure that inventions that have previously been revealed to the public are not granted protection, and in turn, the public is not restricted from benefiting from an invention that already exists.

3.1.3 Inventive step

The requirement of inventive step necessitates that the invention makes a contribution to the state of the art.[17] The steps taken from the prior art to the invention must not be obvious to the person skilled in the art. The patent balance is clearly apparent in this requirement. An invention will not be granted protection if it provides an obvious improvement to previously existing subject matter. Although there have been numerous approaches to determining whether an invention involves an inventive step,[18] the main question

---

[14] For a detailed analysis on the requirement of invention, see: Justine Pila, *The Requirement for an Invention in Patent Law* (OUP 2010).

[15] Article 54 EPC.

[16] Article 54(2) EPC.

[17] Article 56 EPC.

[18] At the EPO, this requirement is determined by the problem and solution approach (EPO Guidelines for Examination, March 2021, G.VII.5). In the UK, the inventive step requirement is determined by the *Windsurfing/Pozzoli* test (Paul Cole and Richard Davis, *CIPA Guide to the Patents Acts* (9th ed, Sweet & Maxwell 2020), 3.05). Although different tests operate to determine obviousness, whether the end results

asked by this provision is whether the invention is obvious. The focus here is on the quality of the invention, which adds another layer to the examination and further consideration of the public interest.

3.1.4 Industrial application

An invention must also have industrial application, or in other words, a useful purpose.[19] For many inventions this requirement is very easy to overcome, as most inventions that are new and inventive will have a useful purpose.

However, the industrial application requirement is particularly important when considering inventions in the field of biotechnology and emerging technologies. Advancements in these fields are time and resource intensive. Important developments occur, resulting in new findings. In some situations, the use of particular findings will not be entirely clear. The industrial application requirement operates in these situations to prevent the grant of patents over subject-matter where its use is not certain.[20] This can also prevent overly broad patents being granted in areas of new development and emerging technologies. The public interest in maintaining access within the patent balance is evident here. If patents were granted in such cases, others would be prevented from further innovation and the public would be disadvantaged by this.

3.1.5 Exceptions

There are also multiple exceptions to patentability. A patent will not be granted for an invention the commercial exploitation of which would be contrary to *ordre public* or morality; for a plant or animal variety or essentially biological process for their production (excluding microbiological processes); or for methods of treatment of the human or animal body (excluding products for use in any method).[21] The subject-matter in question could be a new, inventive, industrially applicable invention, but may not be granted patent protection if it falls within one of these exceptions. The justifications behind the exceptions to patentability are based on public policy. Specific consideration as to when patents should and should not be granted based on particular public interests is unmistakable.

---

would differ is questionable as they are mainly used to provide a structure towards answering the same question. See: *Actavis v ICOS*, [93–96].

[19] Article 57 EPC.

[20] This aspect of the requirement was considered in T 0018/09 *Neutrokine/HUMAN GENOME SCIENCES* off 21.10.2009 and *Human Genome Sciences Inc v Eli Lilly and Company* [2011] UKSC 51, where it was held that it is now a requirement that the use of biotechnical inventions is disclosed in the patent application.

[21] Article 53 EPC.

For example, patents should not be granted on inventions where it would be immoral to commercially exploit them. There are also a number of specific morality-based exceptions that were introduced following developments in the biotechnology sector. The Implementing Regulations of the EPC, which incorporate provisions from the EU Biotech Directive,[22] state in Rule 28 that patents will not be granted for inventions that concern processes for

> cloning human beings; processes for modifying the germ line genetic identity of human beings; uses of human embryos for industrial or commercial purposes; processes for modifying the genetic identity of animals which are likely to cause them suffering without any substantial medical benefit to man or animal, and also animals resulting from such processes.[23]

Plant varieties were protected under an alternative system and excluded to avoid dual protection, and animal varieties were similarly intended to be protected under an alternative system.[24] Their exception was therefore also based on avoiding the unfair possibility of dual protection. Essentially biological processes were seen as natural rather than man-made processes.[25] Finally, it was reasoned that methods of treatment should be freely available for doctors/vets to use on patients without the fear that they would be subject to a case of patent infringement.

The public interest is core to these provisions. They provide the opportunity for the public to oppose patent protection on subject-matter that would be detrimental to society, even if the invention in question meets every other requirement for patentability.

3.1.6 Sufficiency

Moving on to procedural requirements, the invention must be sufficiently disclosed in the patent application.[26] This requirement ensures that the details on how to make the invention in question are provided for in a manner that is

---

[22] Directive 98/44EC of the European Parliament and of the Council of 6 July 1998 on the legal protection of biotechnological inventions (Biotech Directive).

[23] Rule 28(1) Implementing Regulations of the EPC.

[24] The International Union for the Protection of New Varieties of Plants 1961 (UPOV). As can be seen from the *Travaux Préparatoires*, national legislatures were 'at liberty to provide for the grant of national patents for new plant varieties' but European patents would not be allowed. This applied also to the patentability of new animal species. Text available at: www.epo.org/law-practice/legal-texts/epc/archive/epc-1973/traveaux.html.

[25] This provision will be discussed in a detailed case study below.

[26] Article 83 EPC.

clear and complete enough for it to be replicated by the person skilled in the art, without undue burden.

The balance between the interests of the inventor and those of the public is undoubtedly apparent within this requirement. Without sufficient disclosure, the public would not be able to make use of the subject-matter at the end of the period of protection. However, there is no requirement that the *best* method is disclosed, only that one method is. Patent applicants will therefore attempt to keep as much information as possible to themselves to remain at a competitive advantage once patent protection expires.

3.1.7 Clarity and amendments

There are additional procedural requirements operating to ensure that the application and the procedures around it are conducted in a fair manner. For example, the claims must be clear and concise so that once the patent expires, the information contained therein is useful to the public.[27] This provision operates in the public interest as it restricts the patent applicant from making their application so complicated that it would be unintelligible to the person skilled in the art. The application must also relate to a single invention, unless multiple inventions are linked by a single concept, so that the patentee is only afforded protection to one invention rather than bundling a number of inventions under one application in order to save time and money.[28] Finally, if the claims are being amended during various proceedings, this cannot be done in a way that adds matter to the application.[29] This can prevent the patentee from being awarded with a broader scope of protection than that originally intended.

3.1.8 Evaluation

It is clear that the European patent system has been constructed in an attempt to balance the interests of inventors with access to inventions. The public interest is inherently considered in most if not all of the requirements for and exceptions to patentability. Excluding patents on particular types of subject-matter, requiring that an invention is new, inventive, capable of industrial application, providing exceptions to patentability in areas concerning public policy, requiring sufficient disclosure – all invoke the public interest. While some of the criteria are more public-focussed than others, they appear to provide a high bar for protection. However, how these criteria are interpreted and implemented in the systems in which they operate can have a profound impact on the balance between private rights and access.

---

27 Article 84 EPC.
28 Article 82 EPC.
29 Article 123 EPC.

3.2 Interpretation

These provisions are subject to interpretation by numerous institutions at numerous stages and levels. If the national route to protection is taken, the patent application will be examined by a national patent office examiner. If a case comes before a national court, national judges, possibly at multiple levels, will interpret the law and make a decision based on the practice in that country. This could occur in relation to the same invention in multiple countries. If the application involves a biotechnical invention, national courts may have to refer a question to the Court of Justice of the European Union (CJEU) if there is uncertainty surrounding the interpretation of the provisions of the Biotech Directive.[30]

If the European route to protection is taken, applications will be examined by a European Patent Office (EPO) examiner. If an opposition is raised before the EPO, the Opposition Division will make further decisions in relation to grant. If that decision is appealed, the case will appear before the Boards of Appeal, and possibly the Enlarged Board of Appeal. If a unitary patent and unified patent court are introduced, this will add further decision-making institutions.

Clearly, there are multiple opportunities for different bodies to interpret European patent law provisions. Although the law in this area has been harmonised to a certain extent and there have been and are attempts for further harmonisation,[31] the provisions can be and have been interpreted in different ways by different countries and different institutions.[32] How the criteria for patentability are interpreted can have a significant impact on the balance that patent law seeks to achieve.

The aforementioned institutions are particularly aware of the patent bargain as it is explicitly noted in multiple cases before the UK Courts and the Boards of Appeal.

In most of the recent relevant decisions of the UK Supreme Court (UKSC), the patent bargain has been explicitly acknowledged. In *Warner Lambert v Mylan* and *Actavis v ICOS* the UKSC cited a quote by Lord Mansfield from 1778 summarising justifications for the patent system:

> The condition of giving encouragement is this: that you must specify upon record your invention in such a way as shall teach an artist, when your term is out, to make

---

[30] Directive 98/44EC of the European Parliament and of the Council of 6 July 1998 on the legal protection of biotechnological inventions (Biotech Directive).

[31] See: Justine Pila, 'The European Patent: An Old and Vexing Problem' (2013) 62(4) ICLQ 917.

[32] See, for example, various national case law in *Improver v Remington*.

> it - and to make it as well by your directions: for then at the end of the term, the public shall have benefit of it. The inventor has the benefit during the term, and the public have the benefit after.[33]

In both cases, the UKSC went on to discuss the patent bargain in more detail. In *Warner Lambert v Mylan*, it was stated:

> Elementary as it is, it is worth reminding oneself at the outset of the juridical basis on which patents are granted, sometimes called the 'patent bargain'. The inventor obtains a monopoly in return for disclosing the invention and dedicating it to the public for use after the monopoly has expired. The principle remains the foundation of modern patent law.[34]

The same description of the patent bargain appeared in *Actavis v ICOS*.[35] It has also arisen in both *Regeneron v Kymab* and *United Wire v Huawei*.[36] In *Regeneron v Kymab* the UKSC considered this in relation to sufficiency – that if the invention had not been sufficiently disclosed, the bargain would not be achieved.[37] Again, in *United Wire v Huawei*, the court's starting point was the patent bargain.[38]

The bargain has also been raised in EPO jurisprudence. The Boards speak of the principle of the patent monopoly corresponding to the technical contribution of the invention and that the requirements of support and sufficiency serve to ensure that principle: 'The general legal principle that the extent of the patent monopoly, as defined by the claims should correspond to the technical contribution to the article in order for it to be supported, or justified.'[39] This general legal principle, first outlined in *EXXON*, has since been cited in numerous cases – particularly as the background to the requirement of sufficiency, but it has also been taken further. In *AGREVO* the Board applied the principle in relation to inventive step:

> It has for long been a generally accepted legal principle that the extent of the patent monopoly should correspond to and be justified by the technical contribution to the

---

33 *Warner Lambert v Generics (UK) Ltd t/a Mylan* [2018] UKSC 56, [17]; *Actavis Group v ICOS Corporation* [2019] UKSC 15, [53].

34 *Warner Lambert v Mylan*, [17].

35 *Actavis v ICOS*, [53].

36 *Regeneron Pharmaceuticals Inc v Kymab Ltd* [2020] UKSC 27; *United Wire International v Huawei Technologies, Huawei Technologies v Conversant Wireless Licensing SÀRL, ZTE Corporation v Conversant Wireless Licensing SÀRL* [2020] UKSC 37.

37 *Regeneron v Kymab*, [23].

38 *United Wire v Huawei*, [2].

39 T 0409/91 *Fuel Oils/EXXON* of 18.3.1993, 3.3.

> art. […] [T]his general legal principle was applied in relation to the extent of the patent protection that was justified by reference to the requirements of articles 83 and 84 EPC, the same legal principle also governs the decision that is required to be made under article 56 EPC, for everything falling within a valid claim has to be inventive.[40]

It is clear from these few examples that decision-making bodies acknowledge the patent bargain and the balance that is at the core of the European patent system. However, it appears that this consideration is a starting point or background to the main analysis. The bargain appears to be something that is considered as baked into the system, and therefore, by analysing the provision in question, be it patentable subject-matter, sufficiency, inventive step and so on, the court considers the bargain itself. This impression was summed up quite neatly by Lord Hodge in *Actavis v ICOS*, where he stated:

> The general principle that the extent of the patent monopoly should correspond to and be justified by the actual technical contribution to the art is thus part of the jurisprudence of both the EPO and the UK courts and, as observed in *Generics v Warner-Lambert* [...] 'the principal conditions of validity, novelty, inventive step, industrial application and sufficiency are all, in one way or another, directed to satisfying the principle thus expressed'. There is therefore a balance or symmetry in patent law.[41]

He went on to say that the 'patent bargain which Lord Mansfield described and the EPO has used as an overarching principle […] underpins and creates a symmetry between the various provisions of the 1977 Act which govern the validity of a patent'.[42] This underpinning is then relied upon in order to continue with the decision-making process based on a specific provision with the assumption that the public interest has already been thoroughly considered. It is possible that this restricts the courts from considering public interest aspects in further detail, which may have an impact on the overall decision.

Despite the recognition of the balance in these and many other cases, decisions creating imbalance remain. In a number of cases, particularly those relating to the exclusions and exceptions to patentability, the criteria for patentability are interpreted in a particularly narrow manner, causing a shift in the patent balance, impacting the public interest.[43] The shifting balance and its impact on the public has been clearly apparent in relation to the exception

---

40 T 0939/92 *Triazoles/AGREVO* of 12.9.1995, 2.4.2.

41 *Actavis v ICOS*, [57].

42 *Actavis v ICOS*, [82].

43 For an overview, see: Sigrid Sterckx and Julian Cockbain, *Exclusions from Patentability: How Far Has the European Patent Office Eroded Boundaries?* (CUP 2012).

on the patentability of essentially biological processes. In the next section, this chapter will undertake an examination of the development and interpretation of this provision in order to display the impact that reliance on the baked-in bargain and narrow interpretations of exceptions to patentability can have on the public.

4. PRODUCTS RESULTING FROM ESSENTIALLY BIOLOGICAL PROCESSES

As noted above, essentially biological processes are not patentable. Article 53(b) EPC states that 'European patents shall not be granted in respect of essentially biological processes for the production of plants or animals'. There is a long history attached to this exception and it has been the subject-matter of multiple cases before the Enlarged Board of Appeal.[44] This section will consider the historical background of Article 53(b) EPC, its interpretation in the *Broccoli*, *Tomatoes* and *Peppers* decisions, the issues that have arisen as a result and the response to and impact of these decisions. This examination will provide a clear example of how the balance within the patent bargain can shift between the parties involved, and the resulting impact of this on the public interest.

Article 53(b) EPC was based on Article 2(b) of the Strasbourg Patent Convention (SPC).[45] In an examination of preparatory documents of the Conventions, Justine Pila found that had the framers of the EPC investigated the background of Article 2(b) SPC they would have found that the intention behind it was to avoid unifying 'aspects of law affected by public interest issues', because the drafters believed that they did not have the expertise necessary to decide on the issue and due to divergences in national laws.[46] It appears that instead, the inclusion of this provision in the EPC was considered more so in relation to the creation of a 'common patent code'.[47]

In the *Travaux Préparatoires* in relation to the SPC, it can be seen that the wording of this exception was changed from 'purely biological processes' to

---

[44] On the history of the provision, see: Sigrid Sterckx and Julian Cockbain, *Exclusions from Patentability: How far has the European Patent Office Eroded Boundaries?* (CUP 2012) chapter 6; Justine Pila, 'Article 53(b) EPC: A Challenge to the Novartis Theory of European Patent History' (2009) 72 Modern Law Review 436; and Sigrid Sterckx, 'Is the Non-Patentability of "Essentially Biological Process" under Threat?" (2010) 13(1) The Journal of World Intellectual Property 1, 2–5.

[45] Convention on the Unification of Certain Points of Substantive Law on Patents for Invention 1963 (Strasbourg Patent Convention, SPC).

[46] Pila (2009), 441.

[47] Pila (2009), 441.

'essentially biological processes.' This change is detailed in the documents, where it is stated that:

> The text now specifies that the processes which may be ineligible for patents are essentially (and no longer purely) biological. It was evident that this exclusion should be extended to cover processes which were fundamentally of this type even if, as a secondary feature, 'technical' devices were involved (use of a particular type of instrument in a grafting process, or of a special greenhouse in growing a plant), it being understood that such technical devices may perfectly well be patented themselves, but not the biological process in which they are used.[48]

This wording was also used in the EPC. However, in relatively recent practice, uncertainty arose relating to the meaning of 'essentially' biological processes. This was discussed by the Enlarged Board of Appeal in *Novartis II*, where three possible approaches were raised, but no indication as to which one would be preferred was provided. The *Novartis II* decision was seen as a very narrow interpretation of the exception as it only excluded inventions claiming plant varieties directly, which did not include claims that encompassed plant varieties.[49]

Uncertainty remained and in later case law further confusion emerged relating to the explanation of the term 'essentially' in Rule 26(5) of the Implementing Regulations, which was incorporated from Article 2(2) of the Biotech Directive.[50] The provisions of both state: 'A process for the production of plants or animals is essentially biological if it consists entirely of natural phenomena such as crossing or selection.' As crossing and selection involves human intervention, it was unclear how that could be considered essentially biological and what else might also be included.

4.1 *Broccoli I & Tomatoes I*

This question was referred to the Enlarged Board of Appeal in two cases. The Enlarged Board decided to consolidate these cases and gave a joint decision in G2/07 and G1/08 (*Broccoli I & Tomatoes I*), where it elaborated further on

---

48 *Travaux Préparatoires*, Article 53 EPC, available at: www.epo.org/law-practice/legal-texts/epc/archive/epc-1973/traveaux.html. G1/98 *Transgenic plant/NOVARTIS II* of 20.12.1999, p5. This decision was a very narrow interpretation of Article 53(b) EPC as it allows a claim that does not specify plant varieties but may encompass them.

49 Sigrid Sterckx and Julian Cockbain, 'The Patentability in Europe of Plants Produced by Conventional Plant Breeding Processes: The European Patent Office Enlarged Board of Appeal Cases G-2/12 Tomatoes II/STATE OF ISRAEL and G-2/13 Broccoli II/PLANT BIOSCIENCE' (2015) 37(4) EIPR 193, 193.

50 Implementing Regulations to the Convention on the Grant of European Patents of 5 October 1973 (as amended).

the meaning of 'essentially biological process'.[51] The relevant claims related to methods for breeding broccoli with elevated levels of anti-carcinogenic substance and tomatoes with reduced water content.

The Enlarged Board considered the *Travaux* in its decision and came to the conclusion that the change in terminology described above was intended to limit what was patentable to processes that involved more than routine technical measures, that is, where the 'technical measures used are only means serving to bring about the process for the production of plants which are otherwise based on biological forces'.[52] To be patentable, and thus not an essentially biological process, it was held that the process must involve 'an additional step of a technical nature, which step by itself introduces a trait into the genome or modifies a trait in the genome of the plant produced'.[53] The intention of the provision was considered; however, this was as background to the decision-making process. The Enlarged Board focused on the technical aspects of the case, with little additional consideration of the public interest.

This joint decision somewhat resolved the question concerning what is considered an 'essentially' biological process. There was a mixed response to this decision. Some commentators were uplifted by the decision,[54] but for others uncertainty remained and concerns arose that unnecessary steps of a technical nature would be incorporated in order to bypass the exception.[55] The cases were remitted to the Technical Boards of Appeal for a decision on patentability of the inventions in question. Following the guidance from the Enlarged Board, the patentees amended their claims to remove reference to the processes involved but retained claims on the products resulting from the essentially biological processes (as product-by-process claims). This raised yet another issue on the interpretation of the exception in Article 53(b) EPC. The legislation does not state whether the products resulting from such a process are patentable. As a result, the Technical Board of Appeal once again referred questions to the Enlarged Board of Appeal in both cases, on this occasion relating to the patentability of products that are the direct result of an essentially biological process.

---

51 G2/07 *Broccoli/Plant Bioscience* of 9 December 2010 and G1/08 *Tomatoes/State of Israel* of 9 December 2010. For a detailed case background and analysis of the questions referred, see: Sterckx (2010).

52 *Broccoli I & Tomatoes I*, 66.

53 *Broccoli I & Tomatoes I*, 69.

54 Sterckx and Cockbain (2012), 187.

55 Sven JR Bostyn, 'Resolving the Conundrum of the Patentability of Plants Produced by an Essentially Biological Process: Squaring the Circle?' (2013) 35(7) EIPR 383, 387; and Andreas Hubel, 'Essentials on Essentially Biological Processes for the Production of Plants: Decisions G2/07 (Broccoli case) and G1/08 (Tomatoes case) of the Enlarged Board of Appeal of the EPO' (2011) 33(5) EIPR 328, 329.

4.2 *Broccoli II & Tomatoes II*

In G2/12 and G2/13 (*Broccoli II & Tomatoes II*), the Enlarged Board of Appeal interpreted Article 53(b) EPC quite narrowly and concluded that the products of essentially biological processes are patentable.[56]

The Enlarged Board applied numerous methods of interpretation to arrive at this decision. Considering the historical interpretation, reference was made to the legislator's intention as determined in *Broccoli I & Tomatoes I* and it was pointed out that the *Travaux* were silent on the patentability of resulting products. On this basis, the Board concluded that there was no reason in the *Travaux* to exclude such a product from patentability.[57]

The Board also deliberated over secondary considerations, making clear that these issues related to 'questions of law rather than to economic, social and ethical aspects'.[58] On a dynamic interpretation, where the Board considers developments in the area since the implementation of the EPC, it found no reason that the legislator's intention would be different today. It also discussed the issue of legal erosion.[59] Allowing a patent on the product of an essentially biological process arguably makes the exception in Article 53(b) redundant. Patent protection over a product or product-by-process also encompasses the process used to make the invention. Allowing such patents in this situation thus bypasses the process exception. A number of national courts had taken a similar position, and these were also considered.[60] However, the Board disagreed. Referring again to the legislator's intention, it came to the conclusion that broadening the scope of the exception would not erode the provision but instead would introduce an inconsistency in the EPC because material other than plant varieties could generally be protected.[61]

This decision effectively displays the EPO custom of narrowly interpreting exceptions and exclusions to patentability, as well as the reluctance to consider public interest issues. It was heavily criticised.[62] Such interpretative practices

---

[56] G2/12 *Broccoli II* of 25 March 2015 and G2/13 *Tomatoes II* of 25 March 2015.

[57] *Broccoli II & Tomatoes II*, VII, 5(5) and 6(1).

[58] *Broccoli II & Tomatoes II*, VIII(2).

[59] *Broccoli II & Tomatoes II*, VIII, 2.

[60] For example, as elaborated on in Julian Cockbain and Sigrid Sterckx, 'Are Products of Essentially Biological Processes Patentable in Europe? The Purple Radish Sprouts Case in the Netherlands' (2012) 34(6) EIPR 422, the Dutch courts did not allow a patent on a radish obtained by means of an essentially biological process (*Taste of Nature v Cresco*). Legislation in Germany also provided that products resulting from essentially biological processes would not be patentable.

[61] *Broccoli II & Tomatoes II*, VIII, 2(6)(a).

[62] See: Timo Minssen and Ana Nordberg, 'The Impact of Broccoli II & Tomato II on European Patents in Conventional Breeding, GMO's and Synthetic Biology: The

have an immediate effect on the public and the impact of the interpretation in the case at hand is significant from a societal perspective. Allowing patents on such products has an impact on farmers' rights, crop resistance to climate change and food security. However, these public interest issues were not given the attention required. The Enlarged Board was clearly aware of such aspects, and even went so far as listing those that were documented and raised in *amicus curiae* briefs. Even more clearly, it stated that 'considering such general arguments in the present referrals does not fall under the judicial decision-making powers of the Enlarged Board [...] It is not mandated to engage in legislative policy'.[63] However, it is not restricted from doing so and by not considering such issues, this decision is lacking the holistic interpretation necessary to ensure the balance on which the patent system relies. Further, that the Enlarged Board considered these issues to be policy issues and outside the realm of decision-making highlights the perception that public interest matters are a part of patent law only to the extent that they have been written into legislation and will not be considered otherwise. This approach is constraining and prevents progress on a societal level.

4.3 European Commission Notice and Administrative Council Amendments

Following the response to this decision, the European Commission was called on by Parliament to urgently review certain aspects of the Biotech Directive, including the scope and interpretation of the comparable provision excluding from patentability essentially biological processes for the production of plants or animals (Article 4).[64] A Commission Notice was published in November 2016 which stated that the EU legislator's intention with regard to Article 4 was that products of essentially biological processes would not be patentable – contrary to the decision of the Enlarged Board in *Broccoli II & Tomatoes II*.[65] The Commission did recognise the limitation of its assessment and noted that only the CJEU could provide a legally binding decision on the interpretation of the Biotech Directive, leaving open the possibility for a divergent opinion from the CJEU.

---

grand finale of a juicy patents tale?' (2015) 34(3) Biotechnology Law Report 81, available at https://papers.ssrn.com/sol3/papers.cfm?abstract_id=2607865, 21.

[63] *Broccoli II & Tomatoes II*, VIII, 2(6)(c).

[64] European Parliament resolution of 17 December 2015 on patents and plant breeders' rights (2015/2981(RSP)) (2017/C 399/22).

[65] Commission Notice on certain articles of Directive 98/44/EC of the European Parliament and of the Council on the legal protection of biotechnological inventions (2016/C 411/03).

Subsequent to the publication of the Commission Notice, the Administrative Council of the EPO made the surprising decision to amend the Rules relating to the EPC to indicate that products resulting from essentially biological processes would not be patentable. In June 2017, with a significant majority of the Administrative Council in agreement (35/38), Rule 27 was amended to include reference to the new Rule 28(2), which states: 'Under Article 53(b) EPC, European patents shall not be granted in respect of plants or animals exclusively obtained by means of an essentially biological process.'[66] The legality of this amendment has been doubted by academics, practitioners, industry and governments.[67] The Administrative Council has the competence to amend the Implementing Regulations by three-quarter majority,[68] and to amend the Convention to bring it 'into line with an international treaty relating to patents or European Community legislation relating to patents' by unanimity.[69] The amendment to the Implementing Regulations, inserting Rule 28(2), results in a challenge to the established interpretation of Article 53(b) EPC, or in other words could be considered as an amendment to the Convention. By introducing this provision as an amendment to the Implementing Regulations the Administrative Council bypassed the requirement for unanimity, which has caused much concern in the patent law community.[70]

Nevertheless, proceedings that had been postponed were resumed, including an opposition action against another patent on broccoli, which was successful on the basis of the newly introduced Rule 28(2).[71] Furthermore, a patent for an extremely green blocky pepper was rejected by the Examining Division for the same reason. This resulted in an appeal being filed, triggering another significant chapter in this debate.

---

[66] Exclusion from patentability under Article 53(b) EPC of plants and animals produced by essentially biological processes – amendment of Rules 27(b) and 28 EPC (CA/56/17), 6 June 2017; Decision of the Administrative Council of 29 June 2017 amending Rules 27 and 28 of the Implementing Regulations to the European Patent Convention (CA/D 6/17), 29 June 2017; Update on the patentability of inventions concerning plants and animals under the EPC (CA/PL 3/18), 30 Jan 2018.

[67] This is apparent from the *amicus curiae* briefs filed in response to case G3/19, detailed further below.

[68] Article 33(c) EPC and Article 35(2) EPC.

[69] Article 33(b) EPC and Article 35(3) EPC.

[70] These concerns are clearly apparent in a significant number of *amicus curiae* briefs filed in relation to case G3/19, discussed in detail below.

[71] This decision was appealed and the revocation was overturned in T2840/18 *Broccoli type/SEMINIS* of 19 February 2021 implementing the decision in G3/19 *Pepper (follow-up to 'Tomatoes II' and 'Broccoli II')* of 14 May 2020.

4.4 *Blocky Peppers* and the President's Referral

In T1063/18, the Technical Board of Appeal heard the appeal relating to the extremely green blocky pepper.[72] In the Board's reasoning it was decisively stated that Rule 28(2) conflicted with the decision of the Enlarged Board in *Broccoli II and Tomatoes II*.[73] It was determined that Rule 28(2) had amended the meaning of Article 53(b) EPC as interpreted by the Enlarged Board, that this was beyond the competence of the Administrative Council, and that therefore it should not be taken into account.[74]

The Board referred to Article 164(2) EPC in its decision, which provides that 'in case of conflict between the provisions of the Convention and those of the Implementing Regulations, the provisions of the Convention shall prevail', as well as to decision T39/93, which determined that the meaning of an Article, as interpreted by the Enlarged Board, cannot be overturned by a Rule.[75] As a result, products resulting from essentially biological processes were deemed to be patentable again.

The decision, unsurprisingly, caused a stir. Meetings were held where views were exchanged by the Committee on Patent Law, the EPO, representatives of the 38 EPO Contracting States and the European Commission. Due to the uncertainty generated, it was decided that the President of the EPO would make a referral to the Enlarged Board of Appeal (G3/19).[76] The President has the authority to refer questions on points of law to the Enlarged Board but only in certain specific and restricted circumstances, that is, 'where two Boards of Appeal have given different decisions on that question'.[77] The Technical Board of Appeal in *Peppers* had not decided to refer this case because in its view, there was no conflicting case law.

---

[72] T1063/18 *Extreme dark green, blocky peppers/SYNGENTA* of 5 December 2018 (*Peppers*).

[73] *Peppers*, 46.

[74] *Peppers*, 36.

[75] *Peppers*, 26.

[76] 'Referral of a point of law to the Enlarged Board of Appeal by the President of the European Patent Office (Article 112(1)(b) EPC) – G3/19' available at: http://documents.epo.org/projects/babylon/eponet.nsf/0/09D15FA10C1A3A55C125856C0057B988/$File/Referral%20under%20Art.%20112(1)(b)%20EPC_G%2003-19.pdf (President's Referral). On the uncertainty created, see: Ángela Martínez López, 'Decision T1063/18 of the EPO Technical Board Re-sows the Seed of Uncertainty for the Patentability of Plant Products Derived from Essentially Biological Processes' (2020) 42(1) EIPR 19. Another position taken was that the Board in *Peppers* did not have the competence to make such a decision in the first place: Aloys Hüttermann, 'Who Decides if There Is a Conflict Between Implementing Regulations and Articles of the European Patent Convention' (2019) 14(12) JIPLP 958.

[77] Article 112(1)(b) EPC.

Regarding the conflicting cases, the President argued that the different decisions related to how a conflict between Article 53(b) EPC and a Rule clarifying its meaning is examined under Article 164(2) EPC.[78] By relying on T39/93, the Board of Appeal in *Peppers* was in conflict with G2/07, which did not endorse the approach taken towards Article 164(2) EPC in T39/93.[79] The President also argued that if this argument was not accepted the referral should be admissible by 'analogy' because 'where the impossibility for the Office to bring its examination practice in line with the law results from a conflict between an EPC Rule and a single decision of one Board of Appeal [...] the situation is comparable.'[80] Both lines of argument appear to stretch the bounds of Article 112 EPC to breaking point.

In response to the communication from the Enlarged Board outlining the questions raised and the composition of the Board, and opening the case to third party statements, 41 *amicus curiae* briefs were filed by interested parties.[81] Within these briefs a wide-ranging display of various public interests can be identified, including various opinions from the agritech industry, law firms, intellectual property offices, national governments, activist groups, individuals and more. Opinions were understandably mixed. Many discussed the (in)admissibility of the referral and others ignored this point to focus on the (il)legality of Rule 28(2) EPC and the patentability of products resulting from essentially biological processes.

Notably, the series of events leading up to this created a catch-22 situation for some. Those who believed the introduction of Rule 28(2) EPC was beyond the competence of the Administrative Council, that the President's questions were inadmissible, and simultaneously that products resulting from essentially biological processes should not be patentable were in a no-win situation. Indeed, prior to the Referral, the Chartered Institute of Patent Attorneys (CIPA) released a position paper outlining the ways in which the situation could be resolved in a lawful manner.[82] The possible outcomes included accepting the way in which Article 53(b) EPC was interpreted by case law, an

---

78 President's Referral, 18.

79 President's Referral, 12, 13.

80 President's Referral, 30. For a detailed account of the President's Referral, see: Emanuela Gambini, 'Settling the Patentability of Products Obtained from Essentially Biological Processes: The EPO President's Referral to the Enlarged Board of Appeal in the Aftermath of "the Pepper Case"' (2019) 10 European Journal of Risk Regulation 587, 591–596.

81 The amicus curiae briefs are available at www.epo.org/law-practice/case-law-appeals/eba/number.html.

82 CIPA, 'Patenting of Plants in Europe Position Paper' (2019) available at: www.cipa.org.uk/policy-and-news/briefing-papers/position-paper-on-patenting-of-plants-in-europe/.

amendment of EU law followed by an amendment of the EPC and postponing any further actions until a decision had been made by the CJEU, and did not include a referral to the Enlarged Board. Notwithstanding this, the Referral was made.

Given the serious uncertainty within the community over the admissibility of the referral, the Enlarged Board's acceptance was met with astonishment, as was the decision itself.

4.5 G3/19 *Pepper (follow-up to 'Tomatoes II' and 'Broccoli II')*

In G3/19, the Enlarged Board of Appeal decided that the President's Referral was admissible, that Rule 28(2) EPC should not be disregarded and, contrary to its own precedent in *Broccoli II & Tomatoes II*, that products resulting from essentially biological processes are not patentable (again).[83] The Enlarged Board considered this question to be of fundamental importance for a number of reasons, none of which included the public interest in whether such products should be patentable.

Before approaching the question of admissibility, the Board re-phrased the questions referred. The first question related to whether the Administrative Council could adopt or amend Rules that interpret an Article in a way that differs from a previous interpretation. The second had asked whether the exception in Rule 28(2) EPC was in conformity with Article 53(b) EPC. The Enlarged Board made the questions more specific to avoid giving the Administrative Council 'carte blanche to deviate from established case law and give a particular meaning to any Article of the EPC by means of the Rules of the Implementing Regulations', to remove the opinion of the President from the second question and to place the focus of the question on the intention of the referral in the first place – to determine, following developments in the area, whether products resulting from essentially biological processes were patentable.[84] The questions were re-phrased as follows:

> Taking into account developments that occurred after a decision by the Enlarged Board of Appeal giving an interpretation of the scope of the exception to patentability of essentially biological processes for the production of plants or animals in Article 53(b) EPC, could this exception have a negative effect on the allowability of product claims or product-by-process claims directed to plants, plant material or animals, if the claimed product is exclusively obtained by means of an essentially

---

[83] G3/19 *Pepper (follow-up to 'Tomatoes II' and 'Broccoli II')* of 14 May 2020 (G3/19).

[84] G3/19, II.5.

biological process or if the claimed process feature define an essentially biological process?[85]

Moving forward on that basis, on the admissibility of the referral the Enlarged Board discussed different cases where the Boards of Appeal have determined whether a Rule can have an impact on a previous interpretation of an Article. Earlier case law (including T272/95) had decided that the Rules can have an impact on the interpretation of an Article, whereas the *Peppers* case ruled that they could not.[86] As a result there were two different decisions from two Boards of Appeal, which was enough to fulfil the requirements of Article 112 EPC.

On the patentability of products resulting from essentially biological processes, the Enlarged Board discussed the usual methods of interpretation (grammatical, systematic, teleological, historical) and used a dynamic interpretation to come to the conclusion that such products were not patentable. A dynamic interpretation allows for the consideration that previous interpretations of provisions are not 'carved in stone', but are subject to later developments if their former interpretation would conflict with the legislator's intention.[87] The Enlarged Board endorsed the conclusions arrived at in *Broccoli II & Tomatoes II*, but acknowledged that Article 53(b) did not prohibit a broader understanding of the exception either.[88] It was recognised that with the introduction of Rule 28(2) EPC, the legal and factual situation underlying *Broccoli II & Tomatoes II* had substantially changed.[89] Considering that the majority of Contracting States had voted in favour of the implementation of Rule 28(2) EPC, and that many did not allow for patents on such products already, the Enlarged Board concluded that this displayed a clear legislative intent against the patentability of products resulting from essentially biological processes, which called for a dynamic interpretation of Article 53(b) EPC.[90] The approach taken in *Broccoli II & Tomatoes II* was therefore abandoned.[91]

The decision has been strongly criticised, mainly in relation to the competence of the Administrative Council in implementing Rule 28(2) EPC in the first place and the Enlarged Board's minimal consideration of the point. Many

---

85 G3/19, III.3.

86 G3/19, VI.5.1, 5.2. This approach has been considered 'unusual' because the earlier case law maintained the approach of the Boards rather than completely changing it, as elaborated in: Matthieu Dhenne, 'G3/19: Peppers that Remain across the Throat' (2021) 43(2) EIPR 119, 120.

87 G3/19, XX, XXI, XXII.

88 G3/19, XXXVI.3.

89 G3/19, XXXVI.3, XXVI.4.

90 G3/19, XXVI.7.

91 G3/19, XXVI.8.

of the *amicus* briefs raised the issue of the separation of powers between the Boards of Appeal of the EPO and the Administrative Council in this regard. The Enlarged Board found that there was no foundation for this argument given the 'particular structural organisation and constitution of the EPOrg'.[92] Many have perceived this decision as a political one, without legal basis.[93] Indeed, the changes introduced by Rule 28(2) EPC completely contradict previous interpretations of Article 53(b) EPC, a change that should have only been achieved through unanimity.

This decision is clearly notable for numerous reasons. Despite the continued debate over the legality of the Administrative Council's actions, a decision has been made. For the purposes of this chapter, what is highly relevant is the atypical decision of the Enlarged Board to endorse a broad reading of the exception. In doing so, it has somewhat re-balanced the situation caused by *Broccoli II & Tomatoes II*. That products resulting from essentially biological processes are not patentable is a success from a societal perspective. This decision shifts the balance back towards the public interest in access and will likely bring about multiple benefits in relation to farmers' rights, crop development and food security.

Although this broad interpretation and the end result is beneficial to society, it is a rare exception to the norm, and it will always be marred by political controversy.

4.6 Evaluation

An examination of European case law and other developments in relation to Article 53(b) EPC provides clearly perceptible examples of a shifting balance.

In earlier case law, a narrow interpretation of the exception is applied and the resulting impact on access leaves the relevant public at risk. Allowing patents on products of essentially biological processes leads to a situation of legal erosion, an erosion of the balance that the provision seeks to maintain. Even considering this aspect of the public interest is rejected by the Boards as something that is beyond them. However, given the presence of the balance within the system itself, it is something that ought to, and needs to, be considered in more depth.

In most recent case law there has been a complete reversal of the position taken on Article 53(b) EPC. This is a significant departure from precedent. However, this rare example of a broader interpretation of an exception to patentability by the Enlarged Board cannot be put down to a consideration of

---

92 G3/19, XXV.3.
93 Dhenne (2021) 120.

the public interest. This aspect of the case was not even discussed in G3/19 and the decision itself is tarnished by the surrounding political controversy from which it cannot be detached. Although the result of this decision is welcome, the manner in which it was achieved is regrettable.

Responses in practice will tell another story. G3/19 was relevant to more than 250 cases before the Examining Division, Opposition Division or Boards of Appeal and previous decisions are currently being amended as a result.[94] Drafting practices will have to change, as well as protection practices in relation to patent and plant variety rights.

The potential for a balance in this area exists. A suggestion was made in T297/88 that a referral to the Enlarged Board to re-examine the interpretation of a provision should be possible if it were in the public interest to do so given developments of a legal or technical nature.[95] Sharing some similarities with the dynamic approach towards interpretation, G3/19 raises the possibility that this approach could be used more often in order to take developments into account. This could be essential in relation to developments in existing technologies and emerging technologies. However, without explicit consideration and acknowledgement of the broader public interest, progress remains questionable.

5. INVOKING THE PUBLIC INTEREST

The points on narrow interpretations having an impact on the patent balance raised by the evaluation of the interpretation of Article 53(b) EPC are by no means exclusive to that provision. They are particularly apparent in the evaluation of most of the exceptions and exclusions to patentability. This practice has a significant impact on the balance between private rights and access, often tilting it in favour of the patentee.

As mentioned above, the public policy and morality exception is one of the core means by which the public interest can be invoked. However, when it is raised, the results are mixed. In *Novartis II*, the provision was narrowly interpreted. Even when the public interest is core to a decision, Boards appear to shy away from making determinations on this basis. For example, it was stated that:

> the objections to patentability submitted by Greenpeace under Article 53(a) EPC fall outside the scope of the referred questions. The Board recognizes that these objections raise questions which are of interest to the public [...] The EPO has not been vested with the task of taking into account the economic effects of the grant

---

94 T2840/18 *Broccoli type/SEMINIS* of 19 February 2021.

95 T297/88 *Nimodipin II* of 5 December 1989, 2.4.iii.

> of patents in specific areas and of restricting the field of patentable subject-matter accordingly.[96]

This attitude was also apparent in relation to Article 53(b) EPC. Given the wording of the public policy and morality provision, the focus is directed to the 'publication or exploitation of the invention', which is a useful way for Boards to diverge from considerations of the invention itself.[97] The Guidelines for Examination also state: 'This provision is likely to be invoked only in rare and extreme cases. A fair test to apply is to consider whether it is probable that the public in general would regard the invention as so abhorrent that the grant of patent rights would be inconceivable.'[98] This raises a significantly high bar, particularly for emerging technologies, where many relevant factors are unknown. If the public policy and morality provision is applicable in only very rare circumstances, there is the possibility of expanding its scope.[99] However, consideration of the public interest as a separate concept could also be taken further and applied more broadly.

An additional option is to introduce an open and clear provision to invoke the public interest in relation to any stage of the patent life cycle. As well as having the public interest baked into legislation, a separate provision requiring its consideration, especially in the case of emerging technologies, would add an additional layer of societal considerations. Criteria for patentability are often used as policy levers.[100] This is an essential feature of the patent system and assists in maintaining a balance between private rights and access. However, this is subject to individual judicial discretion. When there is a reluctance to consider societal interests in the first place, there is no lever to turn. Introducing a public interest provision (or lever) that would require a more detailed consideration of the public interest as it relates to the balance between private rights and access may result in a patent system that considers the societal impact of patents more substantially. Ideally, particularly for emerging technologies, this lever could

---

96 G1/98 *Transgenic plant/NOVARTIS II* of 20 December 1999, 3.9.

97 T1213/05 *Breast and ovarian cancer/University of Utah* of 27 September 2007, 53.

98 EPO Guidelines for Examination, March 2021, G.II.4.1.

99 For more on expanding the scope of this provision, see: Karen Walsh and Naomi Hawkins, 'Expanding the Role of Morality and Public Policy in European Patent Law' in Paul Torremans (ed), *Intellectual Property Law and Human Rights* (4th ed, Wolters Kluwer 2020); and Justine Pila, 'Adapting the *Ordre Public* and Morality Exclusion of European Patent Law to Accommodate Emerging Technologies' (2020) 38 Nature Biotechnology 555.

100 Dan L Burk and Mark A Lemley, 'Policy Levers in Patent Law' (2003) 89(7) Virginia Law Review 1575.

be used at any stage of the patent life cycle to ensure that uses of inventions revealed later can also be taken into account.

Furthermore, where such matters are concerned, this cannot be an individual decision. If such an issue is raised it requires the introduction of an external panel with representatives of all interest groups, including society representatives. Public interest issues need to be central to discussions, and especially a consideration of the possible detrimental impact that granting the patent would have on society and the various publics and interests that encapsulates.

6. CONCLUSION

Despite its indefinable nature, the public interest forms a significant part of the European patent system in force today. It is inherent in the balance that the patent system seeks to achieve between private rights and access to innovation. This balance is attempted through the terms of the patent bargain: the bargain that society is restricted from accessing a protected invention for a certain period of time in order to reward the inventor and encourage further innovation.

The public interest is present on both sides of this bargain, and it is clear from section 3 of this chapter that balancing the public interest was a key factor in the development of European patent legislation. However, the practice of narrow interpretations resulting in an increased grant of patents has shifted the balance towards the most powerful patentees and away from those in need of access. The public interest is baked into European patent legislation, and as such it does not appear to be given much additional consideration. The legislation only provides the opportunity for the public interest to be invoked in very particular circumstances, such as the public policy and morality exception, which itself will only be successful in rare and extreme circumstances. This also limits the interpretation of these provisions by the courts.

The examination undertaken in section 4 of this chapter displayed a situation in which the balance was tilted in favour of patentees and subsequently re-balanced by later case law. However, this is a very rare reversal that is surrounded by political controversy. Typically, judicial bodies shy away from considerations of a societal nature. However, without more in-depth considerations of the public interest, patentability has been extended into areas previously thought to be against such interest. As a result, it is being diluted and its presence within the patent balance is being diminished. With recognised technologies there have been attempted erosions of previously embedded principles that were anchored by the public interest. With emerging technologies, consideration of the public interest is particularly important in order to avoid overly broad patents and excessive patents being granted in areas that may not yet be fully developed.

It is imperative that the public interest is at the forefront of decision-making for all inventions so that the balance created by the patent system no longer disproportionately benefits the most powerful patentees. In order to encourage this practice, this chapter has suggested the introduction of a mechanism by which the public interest can be invoked at any stage of the patent life cycle. A true balance between private rights and access in a way that elevates the public interest is crucial to delivering a more equitable patent system. When power is shifting towards the most powerful patentees and away from public access, this imbalance needs to be assessed and addressed.

7. Repositioning the access to healthcare debate: FRAND and the Internet of Medical Things

Alison Slade

1. INTRODUCTION

As already seen within this edition, emerging biotechnical innovation offers significant promise for public health. The attainment of these health benefits is increasingly being tied to the realisation of digital connectivity as part of the much lauded Internet of Things (IoT).[1] This marriage of network technologies with medical products and services has important social implications, including the provision of more affordable and accessible healthcare solutions. However, it is becoming increasingly evident that the promise of the IoT, including connected medical technologies, may be frustrated by the exertion of patent rights covering inventions included as part of a network 'standard'. This gives rise to many of the same 'access' debates over patented technology in the pharmaceutical industry, which are now repositioned within a new technological setting. Issues such as affordability, geographical scope and the unwillingness to develop new and promising treatments (as cost of production is seen to outweigh potential economic return) appear equally to apply in this new context. An obligatory commitment to license on fair, reasonable and non-discriminatory terms (FRAND) is the current solution to this patent law obstacle. However, much uncertainty surrounds the nature and scope of FRAND, and little if any attention has been directed at whether the FRAND licensing commitment can accommodate the broader public interest objectives

[1] Deloitte, 'MedTech and the Internet of Medical Things: How Connected Medical Devices are Transforming Healthcare' (2018) 10 www2.deloitte.com/global/en/pages/life-sciences-and-healthcare/articles/medtech-internet-of-medical-things.html accessed August 2021; World Health Organisation, Global Strategy on Digital Health 2020-2025 (2021) www.who.int/docs/default-source/documents/gs4dhdaa2a9f352b0445bafbc79ca799dce4d.pdf accessed August 2021.

to which medical technologies give rise. It is this gap in the literature that this chapter seeks to address.

Simply put, the IoT describes networks of devices that are connected to the internet and communicate and share data with other devices and systems. The 'thing' in IoT denotes any object or physical entity that has sensors, software and connecting technologies attached or embedded from which data about its use and environment is tracked and communicated. Almost any object can be appropriately 'connected' and form part of the IoT. Common examples include digital or mechanical machinery (such as automobiles, domestic appliances and computers), energy systems, city infrastructures and even humans.

In terms of the latter, a subcategory of IoT, often termed the Internet of Medical Things (IoMT), is integrating biological systems to, *inter alia*, monitor and control implanted or wearable medical devices, facilitate pharmaceutical delivery and develop preventative, rather than reactive, healthcare strategies. For example, in the context of precision medicine the IoMT can be employed to gather timely health related information about individuals directly from diagnostic devices; and non-invasive pre-natal diagnosis and monitoring is starting to place reliance on IoT digital connectivity to empower families by providing knowledge and control over their reproductive risks.[2]

The benefits promised by the IoMT go far beyond the provision of improved medical treatments. Significantly lower care costs and increased access to healthcare services are key objectives of the IoMT,[3] something traditional healthcare models have often struggled to achieve.[4] However, 'smart' objects or the IoT/IoMT create interesting legal challenges[5] that must be addressed if the same 'access' issues that have historically surrounded many patented pharmaceutical products are to be avoided and their full potential is to be realised. Important among those is the need for manufacturers of connected devices to obtain quick and affordable access to technical standards that provide necessary connectivity to global communication networks.[6] Yet, many of these standards incorporate patented technologies which have been declared *essential* to its operation by the relevant standard setting organisation (SSO). This

---

2 Liping Chen, Bo Fan, Yan Cai & Qi Shi, 'Application of IoT Medical Image Detection and Prenatal Genetic Testing In Obstetric Clinic' (2021) 81 Microprocessors & Microsystems doi.org/10.1016/j.micpro.2020.103705.

3 Deloitte (n 1) 10.

4 See for example, OECD, *Pharmaceutical Innovation and Access to Medicines* (OECD Publishing 2018).

5 Matthew Hunt & Pat Treacy, 'Internet of Things (IoT): Key Legal Issues' (2020) Lexis PSL Practice Note.

6 EU Commission, 'Setting out the EU approach to Standard Essential Patents' COM (2017) 712 final, 1.

fateful blend of market power gained through the standardisation process and proprietary rights acquired through the patent system can create formidable barriers to market access for manufacturers and increase costs to consumers.

It is important to note that formal technical standards, which arise through the actions of a standard setting organisation (SSO), are developed to ensure interoperability and compatibility for all users and competitors. Accordingly, they are designed to add value to networks and the products that utilise them, and to facilitate innovation. In the context of the IoT, technical standards allow 'smart' devices to connect and communicate over the relevant networks, such as 4G, 5G and WiFi, and they simplify the acquisition and deployment of the technology in question. Yet, technical standards are also regularly regarded as the antithesis of technological innovation. This is especially so where the standard is bound up with technology that is subject to patent protection. These 'standard essential patents' (SEPs) are extremely powerful legal tools with the potential to frustrate the emergence of new technologies. Therefore, extracting the full value of the IoT/IoMT may rest upon the ability of the law to prevent SEP holders from opportunistically leveraging the 'essentiality' of their rights. To address this problem, the regulatory policies of SSOs require that all SEP holders disclose the existence of any intellectual property rights (IPRs) relating to a standard, and, importantly for this analysis, give an irrevocable undertaking to license the patent protected technology on FRAND terms. However, the effectiveness of the commitment is uncertain, as evidenced by the volume of litigation on this issue and the extensive commentary that questions, *inter alia*, judicial reasoning, the legal nature of the FRAND commitment and the economic justification for FRAND's very existence.

In the shadow of this uncertainty, this chapter will build upon recent commentary to consider how FRAND licensing commitments undertaken by SEP holders should be interpreted within the new technological paradigm of the IoT. This analysis will directly examine to what extent the FRAND licensing commitment, as a private law obligation, can accommodate broader public interest objectives arising in the IoMT sector, in particular the need for lower costs and wider access to connected healthcare products, and facilitate a level of legal certainty that will encourage innovation.

2. SEPS AND FRAND: UNDERSTANDING THE DYNAMICS

As mentioned, technological standards are integral to efficient and effective interoperability between devices and networks and are thus central to facilitating competition in industry.[7] Their development often involves significant upfront investment in creating new technologies,[8] elements of which may form the subject matter of patent protection. The relationship between technology standards and patents can be described as both mutualistic and parasitic. The mutualistic nature of the relationship arises because both systems seek to support and encourage innovation. SSOs encourage the use and development of innovative solutions to technological problems by engaging R&D participants at the level of the standard.[9] The patent system, and the possibility of holding a patent on technology that is essential to a standard (and thus indispensable to the industry in general), encourages companies to participate and invest in the standard setting process.[10] Yet, this relationship is also parasitic. Where an SEP exists, all companies that wish to make or sell products that comply with the standard must obtain a licence. The opportunity therefore arises for an SEP holder to extract a commercial benefit to the detriment of the standard setting agenda. Patent rights may be leveraged to allow the SEP holder to extract unreasonable royalties, license on discriminatory terms or even prevent competitors from launching rival products.[11] In these situations, the threat of a prohibitory injunction operates to hold an industry to ransom.[12] Future implementers are locked into a patented standard that they find difficult

---

[7] ibid; Group of Experts on Licensing and Valuation of Standard Essential Patents, 'Contribution to the Debate on SEPs' (2021) 22 https://ec.europa.eu/docsroom/documents/45217 accessed August 2021.

[8] Justus Baron & Daniel F Spulber, 'Technology Standards and Standard Setting Organizations: Introduction to the Searle Center Database' (2018) 27 J. Econ. Manage. Strat. 462, 466.

[9] JRC Science and Policy Report, 'Fair, Reasonable and Non-Discriminatory (FRAND) Licensing Terms' (Publications Office of the EU 2015) [2.2] https://ec.europa.eu/jrc/en/publication/eur-scientific-and-technical-research-reports/fair-reasonable-and-non-discriminatory-frand-licensing-terms-research-analysis-controversial accessed August 2021; Knut Blind & Axel Mangelsdorf, 'Motives to Standardize: Empirical Evidence from Germany' (2016) 48 Technovation 13.

[10] Marc Rysman & Timothy Simcoe, 'Patents and the Performance of Voluntary Standard-Setting Organizations' (2008) 54 Management Science 1920, 1933; Josh Lerner & Jean Tirole, 'Standard-Essential Patents' (2015) 123 J.Pol. Econ 547, 548.

[11] Mark A Lemley & David McGowan, 'Could Java Change everything? The Competitive Propriety of a Proprietary Standard' (1998) 43 Antitrust Bulletin 715, 760.

[12] See for example, Mark A Lemley & Carl Shapiro, 'Patent Holdup and Royalty Stacking' (2007) 85 Tex. L. Rev. 1991; Daniel G Swanson & William J Baumol, 'Reasonable and Non-Discriminatory (RAND) Royalties, Standards Selection, and Control of Market Power' (2005) 73 Antitrust L.J. 1; Jorge L Contreras, 'Fixing FRAND: A Pseudo-Pool Approach to Standards-Based Patent Licensing' (2013) 79 Antitrust LJ 47.

or impossible to access, with no workable alternatives that could provide viable entry to the market for new products or services.[13]

As part of the solution, SSOs have adopted regulatory policies that aim to ensure that the rights of intellectual property owners complement rather than undermine the objectives of standardisation. The most important of these is an irrevocable undertaking to license the IPR protected standard technology to implementers on fair, reasonable and non-discriminatory terms – that is, the 'FRAND' commitment. For example, as stated in the IPR policy of the European Telecommunications Standard Institute (ETSI):

> When an ESSENTIAL IPR relating to a particular STANDARD or TECHNICAL SPECIFICATION is brought to the attention of ETSI, the Director-General of ETSI shall immediately request the owner to give within three months an irrevocable undertaking in writing that it is prepared to grant irrevocable licences on fair, reasonable and non-discriminatory ('FRAND') terms and conditions under such IPR to at least the following extent:

- MANUFACTURE, including the right to make or have made customized components and sub-systems to the licensee's own design for use in MANUFACTURE;
- sell, lease, or otherwise dispose of EQUIPMENT so MANUFACTURED;
- repair, use, or operate EQUIPMENT; and
- use METHODS. The above undertaking may be made subject to the condition that those who seek licences agree to reciprocate.[14]

As observed by Anne Layne-Farrar, 'the FRAND commitment can be thought of as a quid pro quo: in exchange for inclusion in a standard, the patent holder commits to granting licenses on FRAND terms and conditions'.[15] However, while FRAND commitments can be comprehensive in terms of the use licensees can make of the patented technology, the nature and scope of the FRAND obligation itself, that is, what defines fair, reasonable and non-discriminatory licensing terms or procedures, is generally not stated. This is especially pertinent in the context of ETSI, which, as a partner in the Third Generation Partnership Project (3GPP), has a key role to play in developing 5G technology

---

[13] Jorge L Contreras, 'A Brief History of FRAND: Analyzing Current Debates in Standard Setting and Antitrust Through a Historical Lens' (2015) 80 Antitrust LJ 39, 42; Anne Layne-Farrar, 'The Economics of FRAND' in Roger D Blair & D Daniel Sokol (eds), *Antitrust, Intellectual Property and High Tech* (CUP 2017) 58, 61.

[14] ETSI, 'Rules of Procedure: Intellectual Property Rights Policy' (2020), paragraph 6.1 https://www.etsi.org/images/files/IPR/etsi-ipr-policy.pdf accessed August 2021.

[15] Anne Layne-Farrar (n 13) 63.

standards,[16] access to which will be important for development of the health-care technologies at the heart of this analysis.

Despite this lack of detail, what SSOs' IPR policies do express, albeit in broad terms, is an overarching objective to achieve a balance between access to SEPs in order to implement the standard, and a fair reward to the patent holder for their contribution. As stated in paragraph 3.1 of ETSI's IPR Policy, 'the ETSI IPR POLICY seeks a balance between the needs of standardization for public use in the field of telecommunications and the rights of the owners of IPRs'.[17] In practical terms, the policy seeks to avoid the dual evils of 'patent hold-up' and 'patent hold-out'. As mentioned above, patent hold-up arises where the patentee seeks to obtain excessive royalty rates or refuses to license, thus 'holding up' the implementation of the standard.[18] The converse problem of patent hold-out arises where the implementing party offers below-FRAND terms or makes no offer at all, while at the same time making infringing use of the SEP. As recently observed by the UK Supreme court in *Unwired Planet v Huawei*:

> [I]mplementers [...] might knowingly infringe the owner's Essential IPRs by using the inventions in products which meet the standard while failing to agree a licence for their use on FRAND terms, including fair, reasonable and non-discriminatory royalties for their use. In circumstances where it may well be difficult for the SEP owner to enforce its rights after the event, implementers might use their economic strength to avoid paying anything to the owner. They may unduly drag out the process of licence negotiation and thereby put the owner to additional cost and effectively force the owner to accept a lower royalty rate than is fair.[19]

The balance between securing access to SEPs and a fair reward for the patentee has manifested itself in a contractual derogation from a patentee's right to obtain an injunction to prevent infringement of the patent, and an obligation on the part of the SEP holder and the implementer to negotiate a licence of FRAND terms.[20] However, while the FRAND commitment significantly limits the ability of the patentee to obtain injunctive relief, it does little in itself to resolve deadlocks that can occur over licensing terms and conditions.

Disagreements between SEP holders and implementers over what FRAND is have led to protracted and costly litigation within the telecommunications

---

[16] ETSI, '3GPP Telecom Management' https://www.etsi.org/technologies/3gpp-telecom-management accessed August 2021.

[17] ETSI (n 14).

[18] ibid paragraph 6.1.

[19] *Unwired Planet International Ltd v Huawei Technologies (UK) Co Ltd* [2020] UKSC 37 [10].

[20] ibid [14].

industry around the world.[21] Without any clear guidelines from SSOs, national and regional courts are left to struggle with determining what FRAND is, leading to inconsistency in application and delays to the introduction of products to the market. In addition, SSOs' IPR policies, including FRAND, were developed at a time when standards were created and implemented by a limited number of companies operating in the same industry, and thus having common licensing practices.[22] This is no longer the case. Digital communication is now vital to a wide variety of complex and integrated technologies, leading to an increase in SEP owners and implementers from across industry. 'As a result, it has become more difficult to identify a consensual interpretation of FRAND licensing principles' that can facilitate quick and affordable access to the SEP in question.[23] Most disruptive in this context is the Internet of Things.

3. THE INTERNET OF THINGS AND THE LIMITATIONS OF FRAND

The Internet of Things results in the digital connectivity of a vast range of physical components and environments, and it is changing the global economy.[24] The IoT defines networks of connected devices known as 'objects' or 'things'. These devices use sensors to collect and communicate data about their use and the surrounding environment. This data can be stored and analysed, forming part of a 'smart' product or service that can be supplied to consumers.[25] According to a 2017 EU Commission Communication, '[t]he estimated economic potential of IoT applications in devices for humans, homes, offices,

---

21 Chryssoula Pentheroudakis & Justus A Baron, 'Licensing Terms of Standard Essential Patents: A Comprehensive Analysis of Cases' (2019) EU Commission, JRC Science for Policy Report, 18 https://ec.europa.eu/jrc/en/publication/eur-scientific-and-technical-research-reports/licensing-terms-standard-essential-patents-comprehensive-analysis-cases accessed August 2021.

22 ibid.

23 Yann Ménière & Nikolaus Thumm, 'Fair, Reasonable and Non-Discriminatory (FRAND) Licensing Terms: Research Analysis of a Controversial Concept' (2015) JRC Science and Policy Report 3 https://publications.jrc.ec.europa.eu/repository/bitstream/JRC96258/jrc96258.pdf accessed 14 May 2021.

24 McKinsey, 'Unlocking the Potential of the Internet of Things' (2015) www.mckinsey.com/business-functions/mckinsey-digital/our-insights/the-internet-of-things-the-value-of-digitizing-the-physical-world# accessed August 2021; World Economic Forum, 'Shaping the Future of the Internet of Things and Urban Transformation' (2018) www.weforum.org/platforms/shaping-the-future-of-the-internet-of-things-and-urban-transformation accessed August 2021.

25 EU Commission, 'Internet of Things' (2019) https://digital-strategy.ec.europa.eu/en/library/internet-things-brochure accessed August 2021.

factories, worksites, retail environments, cities, vehicles and the outdoors will be up to EUR 9 trillion per year by 2025 in developed countries'.[26]

Much of the potential promised by the IoT rests upon the ability of manufacturers to identify and access SEPs to ensure interoperability and connectivity to global communication networks. However, many commentators are now questioning whether the existing legal framework, which developed in the narrow standard-setting context of telecommunications and cellular technologies, can adequately respond to standardisation in the world of 'smart' objects or IoT.[27]

Increased and timely standard setting is a necessary priority in many areas of digital technology, and, in particular, in network connectivity to facilitate innovation.[28] As observed by the EU Commission, '[a]reas such as eHealth, smart energy, intelligent transport systems and connected and automated vehicles, including trains, advanced manufacturing, smart homes and cities and smart farming will significantly benefit from the proposed prioritisation of standards'.[29]

However, as noted above, the nature of SEPs and the uncertainty surrounding FRAND commitments needs to be carefully evaluated in the IoT context. In a more recent Communication, the EU Commission has identified 'five challenges' as requiring attention in order to support the emergence of the IoT.[30] Among those is the need 'to facilitate access to IP (and therefore allow the take up and diffusion of technologies)'.[31] Of particular concern is the need for a 'clearer and more predictable [licensing] framework, incentivising good faith negotiations rather than recourse to litigations'.[32]

---

26 EU Commission (n 6) 1. The Commission quoted these projections from the 2015 report conducted by the McKinsey Global Institute, McKinsey (n 24).

27 Beatriz Conde Gallego & Josef Drexl, 'IoT Connectivity Standards: How Adaptive Is the Current SEP Regulatory Framework?' (2019) 50 IIC 135; Rupprecht Podszun, 'Standard Essential Patents and Antitrust Law in the Age of Strandardisation and the Internet of Things: Shifting Paradigms' (2019) 50 IIC 720; Mathew Heim & Igor Nikolic, 'A FRAND Regime for Dominant Digital Platforms' (2019) 10 JIPITEC 38.

28 Gallego & Drexl (n 27) 136; European Parliament Policy Report, 'Standard Essential Patents and the Internet of Things: In-Depth Analysis' (2019) www.europarl.europa.eu/RegData/etudes/IDAN/2019/608854/IPOL_IDA(2019)608854_EN.pdf accessed August 2021.

29 EU Commission, 'ICT Standardisation Priorities for the Digital Single Market' COM(2016) 176 final [3.1].

30 EU Commission, 'Making the most of the EU's Innovation Potential: An Intellectual Property Action Plan to Support the EU's Recovery and Resilience' COM(2020) 760 final, 2.

31 ibid 3.

32 ibid 3 & 13.

The expansion of telecoms and communications SEPs to other sectors in light of the IoT also introduces a vast range of economic actors into the SEP/FRAND debate.[33] These new actors often lack familiarity with licensing practices in the ICT sector, especially small and medium-sized enterprises who may also lack the resources to deal with FRAND licensing complexities.[34] Moreover, given the relatively recent integration of connected technologies into other sectors, there is limited established practice against which a comparable licensing analysis of similarly situated parties can be made, this being a key source of data in quantifying FRAND terms.[35] Therefore, the current business models and licensing practices of these new actors demand more flexibility and diversity in SEP licences if the IoT revolution is to fully materialise.

Typifying concerns over smooth and transparent access to SEPs for all sectors is the recent litigation in Germany arising in the automotive industry.[36] Patent lawsuits brought against car manufacturers for infringement of SEPs have resulted in complaints regarding the SEP holders' failure to license on FRAND terms and for violations of unfair competition laws. A key factor arising in these disputes is whether the licensing models that have been developed in the context of the telecommunication and information sector are appropriate and acceptable to those operating in other sectors. For example, the Court of Justice of the European Union (CJEU) has recently been asked by the Higher Regional Court of Düsseldorf to provide clarity on, *inter alia*, whether Nokia's refusal to license its SEPs to upstream automotive component suppliers and to seek an injunction against the downstream company, Daimler, for patent infringement is contrary to its FRAND commitments and thus an abuse of a dominant position within the meaning of Article 102 of the Treaty on the Functioning of the European Union (TFEU).[37] It is common practice within the mobile device industry to license at the final stage of the manufacturing – that is, at a price per consumable unit.[38] While the litigation between Nokia and

---

33 Gallego & Drexl (n 27) 137.

34 Group of Experts on Licensing and Valuation of Standard Essential Patents (n 6) 42.

35 Pentheroudakis & Baron (n 21) 14.

36 For an overview of the litigation between Nokia and Daimler see Mathieu Klos, 'All Eyes on Luxembourg in Nokia and Daimler Patent Battle' (*JUVE Patent*, 24 February 2021) www.juve-patent.com/news-and-stories/cases/all-eyes-on-luxembourg-in-nokia-and-daimler-patent-battle/ accessed August 2021.

37 C-182/21 *Nokia Technologies v Daimler AG* (26 November 2020) Request for a preliminary ruling.

38 See, for example, the detailed analysis of licensing practice within the mobile telecommunications sector undertaken by Birss J in the UK High court decision in *Unwired Planet International Ltd v Huawei Technologies Co Ltd* [2017] EWHC 711 (Pat).

Daimler has subsequently been settled, it does illustrate that the licensing of SEPs for the IoT challenges the standardisation and licensing frameworks currently in existence. While royalty rates based upon entire market value may be acceptable in the communications and telecoms sector, the automotive sector, for example, takes a very different approach. As noted by Damien Geradin, the vehicle manufacturing industry has historically required component suppliers to deliver their parts free of third-party rights, including IPRs.[39] The question of whether the FRAND commitment undertaken by SEP owners necessary obliges a 'license to all' approach – thus requiring patent holders to license to all parties interested in acquiring one, irrespective of their level in the supply chain – is one of the pressing questions yet to be resolved.

As the above highlights, the lack of a clear and predictable licensing framework and the emergence of new actors with different visions of what FRAND is give rise to several contentious legal issues that need to be addressed in order to ensure that FRAND facilitates rather than hinders innovation and access in the IoT. However, little, if any, consideration has been given to the wider public interest perspectives that influence business practice in many industries and how they should also be reflected in FRAND licensing practice. As the rest of this chapter will consider, the economic and policy objectives that underpin FRAND determinations in the field of telecommunications fail to adequately reflect the new public policy positioning established by the amalgamation of the communications and life sciences sectors. Consequently, the IoMT necessitates a further reformulation of what is FRAND to ensure that IoMT also achieves the promise of lower costs and increased access to healthcare.

4. FRAND AND THE INTERNET OF MEDICAL THINGS

The growth of medical devices, sensors and services that connect to health networks via digital communication systems, commonly referred to as the Internet of Medical Things (IoMT), is revolutionising healthcare across the world. This transformation has accelerated in light of the COVID-19 pandemic, which has required limits to be placed on travel, imposed social distancing on individual interactions and raised public awareness about the importance of monitoring and maintaining a healthy lifestyle, both physically and mentally. The result has been a growth in demand for connected products and services, and

[39] Damien Geradin, 'SEP Licensing after Two Decades of Legal Wrangling: Some Issues Solved, Many Still to Address' (2020) TILEC Discussion Paper DP 2020-040, 17 https://papers.ssrn.com/sol3/papers.cfm?abstract_id=3547891 accessed August 2021.

a greater willingness to embrace these new technologies on the part of clinicians, patients, governments and industry.[40] The potential of the IoMT extends beyond facilitating greater efficiency and productivity of traditional healthcare structures, to widening its reach to encompass those that are currently excluded or have little access to healthcare services. For example, connected medical devices can diagnose, monitor and manage widespread chronic diseases such as heart disease, asthma and depression for those not present in a healthcare facility, and empower individuals to monitor and treat their own medical conditions, thus reducing the need for further medical intervention.[41]

As has been highlighted in several chapters of this edition, concerns regarding access to healthcare are longstanding and multifaceted.[42] Limited or no access to healthcare provision has historically been viewed as a consequence of numerous factors, including geographical constraints, financial constraints, time constraints and a lack of knowledgeable medical practitioners.[43] Yet, the IoMT has the potential to remove some of the existing barriers to healthcare and provide a more egalitarian provision. To maximise its potential, and to remove inequalities that may result from difficulties in accessing technology hardware or reliable ICT services, many countries have, or are developing, policies aimed at facilitating IoT infrastructure and its implementation in national healthcare delivery. These include developing nations such as Indonesia, Philippines, Malaysia and India.[44] Benefits for citizens in developing countries

---

[40] Deloitte, 'The Future Unmasked: Predicting the Future of Healthcare and Life Sciences in 2025' (2020) www2.deloitte.com/content/dam/Deloitte/pt/Documents/life-sciences-health-care/predictions-2025/LSHC-Predictions-2025.pdf accessed August 2021.

[41] In their 2018 report on the IoMT, the Deloitte Centre for Health Solutions identified seven main ways in which the IoMT was having a positive impact upon healthcare – improved drug management, decreased costs, enhanced patient experience, improved diagnostics and treatment, remote monitoring of chronic diseases, improved disease management, and improved patient outcomes. Deloitte (n 1) 10.

[42] See for example Dianne Nicol & Jane Nielsen, 'Non-Invasive Prenatal Testing and the Resilience of the Patent System' and Aisling McMahon, 'Accounting for Ethical Considerations in the Licensing of Patented Biotechnologies and Health-Related Technologies: A Justification' in Naomi Hawkins (ed.) *Patenting Biotechnological Innovation: Eligibility, Ethics and Public Interest* (Edward Elgar Publishing 2022); Dina Balabanova, Justin Parkhurst, Martin McKee & Barbara McPake, 'Access to Health Care: Taking into Account Health Systems Complexity' (2006) Department for International Development Working Paper www.gov.uk/research-for-development-outputs/access-to-health-care-taking-into-account-health-systems-complexity accessed August 2021.

[43] Owen O'Donnell, 'Access to Health Care in Developing Countries: Breaking Down Demand Side Barriers' (2007) 23 Cad Saude Publica 2820.

[44] Jaimon T Kelly, Katrina L Campbell & Paul Scuffham, 'The Internet of Things: Impact and Implications for Health Care Delivery' (2020) 22 J. Med. Internet Res. e20135, 4. https://www.jmir.org/2020/11/e20135/ accessed August 2021.

have already been realised from IoMT devices that, for example, conveniently monitor glucose levels, allowing individuals greater control over treatment for their diabetes.[45] Yet, alongside issues regarding technology hardware and secure connectivity provision, the promise of the IoMT is only realisable if those who need healthcare devices can access them on fair terms. Patent rights, over pharmaceutical products in particular, have long been seen as a barrier to accessing affordable medicines and achieving one of the key UN Sustainable Development Goals of 'Ensuring healthy lives and promote well-being for all at all ages'.[46] It is important that those same concerns over patent rights do not transition from pharmaceutical products to the IoMT, via SEPs, and thus perpetuate health inequalities across the globe.

It is accepted that analogies drawn with pharmaceutical patents may be viewed by some as too simplistic. For instance, pharmaceuticals are often protected by patents covering a single active ingredient. Yet, many of the multi-component devices grouped under the banner of the IoMT will be subject to a number of intellectual property rights (IPRs), each of which has the potential to be exerted by its owner in a way that may limit access to the product or service in question. In addition, many patent holders are already making voluntary public pledges to, for example, refrain from asserting their patents or seeking injunctive relief for a specific period of time or for the purpose of addressing a pressing healthcare need, such as COVID-19.[47] Nevertheless, concerns regarding access to SEPs are justified. Gaining access to an SEP on appropriate FRAND terms gives rise to additional legal hurdles and costs, and creates commercial uncertainty. Traversing complex licensing issues creates additional expenditure, in terms of both time and money, that at best increases the cost of production and at worst acts as a disincentive to develop the technology.

---

[45] Luis Eduardo P Calliari, Marcio Krakauer, Andre Gustavo Daher Vianna, Yashesvini Ram, Douglas Eugenio Barbieri, Yongjin Xu & Timothy C Dunn, 'Real-world Flash Glucose Monitoring in Brazil: Can Sensors Make a Difference in Diabetes Management in Developing Countries?' (2020) 12 Diabetology & Metabolic Syndrome Journal 1.

[46] United Nations, 'Sustainable Development Goals: Our Framework for COVID-19 Recovery' (2020) www.un.org/sustainabledevelopment/sdgs-framework-for-covid-19-recovery/ accessed 14 May 2021; Siva Thambisetty, 'Improving Access to Patented Medicines: Are Human Rights Getting in the Way?' (2019) 4 IPQ 284, 300.

[47] Voluntary patent pledges have been made by several independent organisations and collaborative networks in the fight against COVID-19. For a brief analysis and overview see Jorge L Contreras, Michael Eisen, Ariel Ganz, Mark Lemley, Jenny Molloy, Diane M Peters & Frank Tietze, 'Pledging Intellectual Property for COVID-19' (2020) 38 Nature Biotechnology 1146.

While the majority of litigation has been confined to the mobile communications sector, the recent litigation between Nokia and Daimler is a clear indicator that the FRAND licensing complexity is set to expand into other sectors of the IoT, including the IoMT. Thus, it is necessary to evaluate whether the broader public interest objectives, such as lower cost and widening access to healthcare, can be accommodated as part of the FRAND analysis which has, to date, been broadly insulated from general welfare-improving considerations.[48] The next section of this chapter will consider in detail the legal nature of the FRAND commitment and whether, in each legal characterisation, principles exist that provide a basis for factoring in health policy objectives in the assessment of what FRAND is.

5. FRAND: A PRIVATE OBLIGATION TO ACHIEVE PUBLIC HEALTH OBJECTIVES

As observed by Jorge Contreras, patent pledges in the form of a FRAND commitment 'are intended to assure the *market*, rather than specific firms, that the pledger's patents will not be used to block adoption of a common technology platform'. As such, they 'preserv[e] the public benefits generated by common technology platforms'.[49] Like the standardisation process and the standards that SEPs are a part of, the wider objective of FRAND is to secure a benefit to the public.[50] Thus, while the FRAND commitment itself is a private ordering mechanism arising from privately organised SSOs, and the widely accepted method of enforcement arises under private contract law remedies, the intended beneficiaries of the FRAND obligation are the public.[51] This public/private interface necessarily creates opportunities as well as challenges for a public health conception of FRAND in the context of the IoMT.

As already discussed, a commitment to license on FRAND terms is a voluntary obligation undertaken by a patent holder to an SSO in order to fulfil the terms of the SSO's IPR policy, which seeks to ensure that IPRs, including

[48] Pentheroudakis & Baron (n 21) 127.

[49] Jorge L Contreras, 'A Market Reliance Theory for FRAND Commitments and Other Patent Pledges' (2015) 2 Utah Law Review 479, 486 & 490.

[50] As noted by Keith Maskus, 'Technical and product standards exist to achieve numerous public policy objectives, such as component interoperability, network reliability, public health, environmental protection, and consumer safety'. Keith E Maskus, *Private Rights and Public Problems: The Global Economics of Intellectual Property in the 21st Century* (Pearson Institute for International Economics 2012) 158.

[51] Jorge L Contreras, 'From Private Ordering to Public Law: The Legal Frameworks Governing Standards-Essential Patents' (2017) 30 Harv. J. L. & Tech. 211, 227.

patents, are not an obstacle to the adoption of standards.[52] The courts in many jurisdictions, such as the UK and the US, tend to treat the obligation as a contractual one, and thus a breach of the FRAND commitment amounts to a breach of contract. Other jurisdictions, including those in parts of mainland Europe, favour a competition law approach, allowing SEP implementers to argue a breach of competition law as a defence to patent infringement. However, the reality is that the FRAND commitment is a complex mix of patent law, contract law and competition law. In short, FRAND is a contractual derogation undertaken by the patent holder with the SSO not to extract the full proprietary value of the patent if designated as an essential part of a technical standard, including not to seek injunctive relief, unless the implementer has failed to conduct licence negotiations in a fair and reasonable manner.[53] As a consequence, FRAND is an obligation to grant a licence on FRAND terms to potential implementers as third-party beneficiaries.[54] A FRAND commitment also creates an obligation under competition law which may be breached should the SEP holder fail to behave in a certain manner, that is, fails to conduct the negotiation of a licence in a FRAND manner.[55] The following will consider in turn each of the legal bases – contract and antitrust – to examine the extent to which – if at all – the wider public health objective of access to healthcare can be factored into the analysis.

5.1 FRAND and contract law

While there is some uncertainty, especially in Europe,[56] the prevailing opinion of many courts and academics is that the FRAND commitment is a contract between the SEP holder and the SSO, with potential implementers of the stand-

[52] It is important to note that the terms of the SSOs' IPR policies can vary considerably, although many adopt a similar 'baseline policy'. Justus Baron, Jorge Contreras, Martin Husovec & Pierre Larouche, 'Making the Rules: The Governance of Standard Development Organisations and Their Policies on Intellectual Property Rights' (2019) Commission, JRC Science for Policy Report, 131–6 https://ec.europa.eu/jrc/en/publication/making-rules-governance-standard-development-organizations-and-their-policies-intellectual-property accessed August 2021.

[53] *Unwired Planet v Huawei* (n 19) [4]–[14].

[54] ibid [8].

[55] C-170/13 *Huawei Technologies v ZTE Corp* [2015] 5 CMLR 14 [71].

[56] Hanns Ullrich, 'FRAND Access to Open Standards and the Patent Exclusivity: Restating the Principles' (2017) Concurrences [31] www.concurrences.com/en/review/issues/no-2-2017/articles/frand-access-to-open-standards-and-the-patent-exclusivity-restating-the accessed August 2021.

ard being viewed as third-party beneficiaries to the contract.[57] Thus, a failure to negotiate a licence to implementers on FRAND terms amounts to a breach of a contractual promise which the implementer can seek to enforce. The contract is necessarily 'incomplete' given the difficulties in identifying all possible contractual contingencies in all situations,[58] and the potential for the SSOs themselves to be guilty of antitrust behaviour should they seek to predetermine or restrict the royalty payable.[59] The incompleteness of the contract signifies that the FRAND commitment does not guarantee the formation of a licensing agreement and is merely a duty to confine the terms to what may be considered 'fair and reasonable' in the circumstances. The question is whether this 'incompleteness' and/or the inherent flexibility in determinations of 'fair and reasonable' provide sufficient room for incorporating into the analysis wider public health objectives.

The very nature of a contract places a significant limitation on this prospect. As directly observed by Roger Brooks and Damien Geradin, if FRAND is a contractual arrangement then 'both the Civil Law and Common Law traditions of contract interpretation and enforcement fundamentally look to discern and give effect to *the intent of the parties*' and not general public policy objectives.[60] Hence it is probably futile, certainly in this context, to try to argue that specific methodologies should be adopted in order to, for example, restrict the royalties payable in order to facilitate greater public access to healthcare devices.[61] Nevertheless, in determining the intent of the parties reference may be had to categories of information that may indirectly incorporate these concerns, such as the industry practice of not just the SEP holder but also the beneficiary of the contract, that is, the industry practice of the implementer.

A detailed examination of licensing practice within the medical technology sector is beyond the scope of this chapter. However, the need to tailor the terms to the parties in question, and the benefits of doing so, were identified by the

---

57 *Apple, Inc. v Motorola Mobility, Inc.*, 886 F. Supp. 2d 1061, 1083-85 (W.D. Wis. 2012); *Unwired Planet v Huawei* (n 19) [8]; Geradin (n 39) 6.

58 Pentheroudakis & Baron (n 21) 33.

59 Mark Lemley, 'Intellectual Property Rights and Standard-Setting Organizations' (2002) 90 California Law Review 1889, 1939–1947. Under Article 101 TFEU, the practice of predetermining licensing terms may be considered anticompetitive.

60 Roger G Brooks & Damien Geradin, 'Interpreting and Enforcing the Voluntary FRAND Commitment' in Kai Jakobs (ed), *Innovations in Organizational IP Specification and Standards* (IGI Global 2013) 52, 55.

61 ibid.

EU Commission in its 2017 communication setting out its approach to SEPs. The Commission observed:

> There is no one-size-fit-all solution on what FRAND is: what can be considered fair and reasonable can differ from sector to sector and over time. Efficiency considerations, reasonable licence fee expectations on both sides, the facilitation of the uptake by implementers to promote wide diffusion of the standard should be taken into account.[62]

However, what the 'expectations on both sides' are and what terms are required to facilitate 'uptake by implementers' to 'promote wide diffusion of the standard' in the context of the IoMT are far from clear at the present time. There are some obvious problems arising from current FRAND formulations which may hinder the potential of the IoMT to meet these objectives and to also provide access to affordable healthcare. The following interrelated questions shall be examined in this context: Is the use of end-product pricing an appropriate royalty base? Does the 'non-discrimination' limb prevent licensees from the IoMT sector being treated differently to those in the mobile communications sector? Do global licensing arrangements negatively impact upon variable pricing for healthcare products?

5.1.1 Is end-product pricing an appropriate royalty base?

The focus of the FRAND analysis is on 'sector' licensing practices. To date, most of the case law, and thus judicial guidance, has been confined to the mobile telecommunications sector. There are, therefore, many questions surrounding the nature of the FRAND licensing terms and how they will be calculated in sectors where connectivity is not the primary objective of the device. Key among those is what should be the appropriate royalty base, that is, the base against which the licence fee will be calculated. As mentioned above, judicial determinations of FRAND royalties have, due to mobile communications sector practice, been linked to the cost of the product to the end user.[63] While the litigation in *Nokia v Daimler* would have seen a review of this practice by the CJEU, the parties' out of court settlement means that this issue will not now be considered and uncertainty remains on the impact this

---

[62] EU Commission (n 6) 8.

[63] Jorge Padilla & Koren W Wong-Ervin, 'Portfolio Licensing to Makers of Downstream End-User Devices: Analyzing Refusals to license FRAND-Assured Standard-Essential Patents at the Component Level' (2017) 62 Antitrust Bulletin 495, 501. The practice of licensing SEPs at end user products has also been prevalent in standards relating to audio and video technologies. Group of Experts on Licensing and Valuation of Standard Essential Patents (n 7) 77.

practice has on royalties payable in the IoT context.[64] It is widely accepted that FRAND terms should be based upon the value of the technology in question. The EU Commission has stressed that the decided value of the SEP 'should not include any element resulting from the decision to include the technology in the standard', and that the added value brought by the SEP should be evaluated 'irrespective of the market success of the product [in which it is incorporated and] which is unrelated to the patented technology'.[65] Those operating in the IoT/IoMT sector rightly argue that licensing of SEPs in this context should reflect the connectivity value the SEP provides, not the price of the end product, which encompasses technology that the SEP holder has contributed nothing towards. As observed by the SEP patent pool, Avanci,

> [I]t is difficult to derive the value of the licensed wireless technology from the sales price of the device since the products have primary uses unrelated to connectivity. For example, the value of a license to wireless technology bears little relation to the $75,000 price tag for a luxury sports car or a $2,000 stainless steel refrigerator.[66]

Thus, continuing to calculate the royalty as a percentage of the entire market value is difficult to justify where the patented technology is not the only driver of demand for the product.

An alternative is that FRAND is perceived as a 'license-to-all' rather than an 'access-to-all' obligation. The distinction is important and speaks to the very nature of the contractual obligation undertaken by the SEP holder. In practical terms, a 'license-to-all' undertaking requires the patentee to grant licences to all potential licensees irrespective of their position in the supply chain. This approach ultimately seeks to contain SEP licensing to the upstream component, that is, the chipset, and, in doing so, potentially reduce the amount of royalties payable.[67] The lower the per-unit price or profit on which the percentage payable is calculated, the lower the royalty paid.[68] The possibility of

---

[64] Anne Layne-Farrar & Richard J Stark, 'License to All or Access to All? A Law and Economics Assessment of Standard Development Organizations' Licensing Rules' (2020) 88 Geo. Wash. L. Rev. 1307, 1348.

[65] EU Commission (n 6) 6–8.

[66] Avanci, 'Accelerating IoT Connectivity' (2016) 6. www.avanci.com/wp-content/uploads/2020/03/Avanci-White-Paper.pdf accessed August 2021.

[67] As noted by Renato Nazzini, while in theory the royalty base should make no difference to the total royalty calculated, 'in practice the larger the royalty base the larger the royalty may actually be'. Renato Nazzini, 'Level Discrimination and FRAND Commitments under EU Competition Law' (2017) 40 World Competition 213, 217; The proposal for a 'principle of neutrality', that is, 'FRAND is FRAND regardless of the licensing level', was made by the Group of Experts on Licensing and Valuation of Standard Essential Patents (n 7) 84.

[68] Layne-Farrar & Stark (n 64) 1309.

the SEP holder 'double dipping', that is, seeking to license at multiple stages of the supply chain, is limited by the doctrine of patent exhaustion.[69] In basic terms, this doctrine holds that the rights of the patent holder are 'exhausted' on first sale of an article that embodies the patented technology, leaving subsequent users, be they manufacturers further down the supply chain or the ultimate consumer, free to use that article free from any claim by the patent holder.[70] Thus, the combination of a 'license-to-all' approach coupled with the exhaustion doctrine would curtail the SEP holder's ability to seek more extensive royalties from those using the technology that embodies the SEP, that is, the chipset, further down the supply chain.[71]

In contrast, the 'access-to-all' argument maintains that the SEP holder should be free to choose where to license in the supply chain, and that the FRAND commitment merely ensures that implementers have 'access' to the technology in question. This approach is founded on the assumption that, when drafting their IPR policies, SSOs 'used language reflecting the prevailing industry practice, i.e. language focused on the manufacture of *equipment* rather than components'.[72] Anne Layne-Farrar and Richard Stark, among others, have observed that the language of many SSOs' IPR policies is generally unclear and inconsistent on this issue.[73] In addition, some SSOs, while incorporating a license-to-all commitment, allow SEP holders to modify that commitment and individually decide where in the supply chain they are willing to license.[74] However, while the access-to-all approach currently appears entrenched, the use of standard technologies to produce devices that go way beyond the mere

---

69 'Double dipping' is a term used to describe a situation where the SEP holder is able to obtain a licence at various levels of the supply chain, thus being over-compensated for the contribution made by the SEP. Group of Experts on Licensing and Valuation of Standard Essential Patents (n 7) 84.

70 For a discussion of the doctrine of exhaustion and the possible incentive it may create in SEP/FRAND situations for the patentee to argue that the patent is not embodied in a particular component and thus not exhausted, but that use of the component in combination with other technology (i.e. use further down the supply chain) now infringes the SEP, see Herbert Hovenkamp, 'Reasonable Patent Exhaustion' (2018) 35 Yale Journal on Regulation 513, 539.

71 For a more detailed examination of the complexities and options involved in the 'license-to-all' model see Group of Experts on Licensing and Valuation of Standard Essential Patents (n 7) 80–5.

72 Bertram Huber, 'Why the ETSI IPR Policy Does Not and Has Never Required Compulsory "License to All": A Rebuttal to Karl Heinz Rosenbrock' (2017) 5 https://papers.ssrn.com/sol3/papers.cfm?abstract_id=3038447 accessed 14 May 2021.

73 Layne-Farrar & Stark (n 64) 1325–7; Jean-Sébastian Borghetti, Igor Nikolic & Nicolas Petit, 'FRAND Licensing Levels under EU Law' (2021) European Competition Journal, DOI: 10.1080/17441056.2020.1862542 accessed August 2021.

74 Layne-Farrar & Stark (n 64) 1325–7.

provision of voice and data services to include *inter alia* health-improving or saving technology may force a review of this practice by new implementers.

5.1.2 Does the 'non-discrimination' limb prevent differential treatment?

While the 'non-discrimination' limb of the FRAND commitment has been held to be 'general' and not 'hard-edged' (that is, it does not require that all similarly situated licensees are entitled to the same terms), the UK Supreme Court has emphasised:

> The 'non-discriminatory' part of the phrase [...] indicates that the terms and conditions on offer should be such as are generally available at a fair market price for any market participant, to reflect the true value of the SEPs to which the licence relates and without adjustment depending on the individual characteristics of a particular market participant. Put another way, there is to be a single royalty price list available to all.[75]

According to the Court, 'non-discrimination' in the FRAND context does not create a separate obligation distinct from 'fair and reasonable'.[76] It 'gives colour to the whole and [...] provides focus and narrows down [...] what might count as "fair" or "reasonable"'.[77] Non-discrimination favours uniform licensing, but not to the extent that certain factors about the licensor and/or licensee are incapable of influencing the terms agreed.[78] As noted in the IPR Policy Guidelines of the Telecommunications Industry Association, 'discrimination and difference are not the same'.[79]

In economic terms, differential or even discriminatory pricing is not necessarily harmful.[80] Jorge Contreras and Anne Layne-Farrar observe that discriminatory pricing can actually raise total welfare as 'more consumers can buy at a price they are willing [and able] to pay' and 'the seller sells more products, and even though profits are lower on sales to the constrained group, they are non-zero thereby increasing overall profits'.[81] It should be no surprise that differential pricing is common in patent licensing. Within the life science sector,

---

75 *Unwired Planet v Huawei* (n 19) [114].

76 ibid [113].

77 ibid [114].

78 EU Commission (n 6) 7.

79 Telecommunications Industry Association, 'Guidelines to the Telecommunications Industry Association Intellectual Property Rights Policy' (2014) 7 https://tiaonline.org/what-we-do/standards/procedures-guidelines/ accessed August 2021.

80 Jorge L Contreras & Anne Layne-Farrar, 'Non-Discrimination and FRAND Commitments' in Jorge L Contreras (ed), *The Cambridge Handbook of Technical Standardization Law* (CUP 2017) 191.

81 ibid.

different end-product markets and consumer purchasing power influence the terms of patent licensing agreements.[82] Accordingly, there is a prima facie case that differential pricing of SEPs to those producing devices that have important global healthcare benefits may be entitled to more favourable licensing terms without the SEP holder being in breach of their FRAND commitment. Many commentators agree that the non-discrimination limb requires less than a 'most favoured licensee' clause,[83] but is more than a reaffirmation of competition law principles which may permit clear discrimination in certain circumstances.[84] However, clear identification of what room exists for differential terms and the factors that justify those terms is currently an open question. While the non-discrimination commitment may provide room for differential terms, it does not, in itself, obligate the SEP holder to do so. It is here that the elements of 'fair and reasonable' may come into play to actually demand different terms in different sectors. What is discriminatory and thus not 'fair and reasonable' in the mobile communications context may actually prove to be FRAND in different sectors. In fact, some have argued that the non-discrimination limb obliges the SEP holder to license to all willing licensees irrespective of their position in the supply chain (and thus potentially reduce the cost of licensing SEPs). When analysis looks to establish what is 'fair and reasonable' in the healthcare context, non-discrimination between licensing practice in different

---

82 For example, the UN backed Medicines Patent Pool has successfully negotiated 22 sublicenses between generic manufacturers and product developers that has allowed citizens in 141 countries to benefit from cheaper and more efficient access to certain pharmaceuticals – 'MPP in Numbers' https://medicinespatentpool.org/progress-achievements/impact/ accessed August 2021. Many pharmaceutical manufacturers also directly embed 'access to medicine' into their general business practices and supply patented products at reduced prices in many developing nations. For a recent report on how 20 of the world's leading pharmaceutical companies are achieving greater access see Access to Medicine Foundation, 'Access to Medicines Index 2021' https://accesstomedicinefoundation.org/publications/2021-access-to-medicine-index accessed August 2021; see also the discussion of the 'tiered pricing' structure adopted by many pharmaceutical companies in David Henry & Andrew Searles, 'Pharmaceutical Pricing Policy' in Management Sciences for Health Inc, *Managing Access to Medicines and Health Technologies* (2012) [9.4] www.msh.org/sites/default/files/mds3-jan2014.pdf accessed August 2021.

83 In general terms, a most-favoured licensee clause is one which guarantees that *all* licensees are entitled to the same preferential terms granted provided to another licensee. While clearly being non-discriminatory, their impact can be to keep terms and pricing artificially high to avoid a race to the bottom: Contreras & Layne-Farrar (n 80) 199.

84 Brooks & Geradin (n 60) 67; Jorge L Contreras & Richard J Gilbert, 'Non-discrimination: FRAND's Last Stand?' (2020) Competition Policy International https://www.competitionpolicyinternational.com/non-discrimination-frands-last-stand/ accessed August 2021.

sectors may prove to be less of an obstacle and provide the required flexibility to accommodate variation in consumer purchasing power in different global markets. Yet, the other common practice of licensing on a global basis also needs to be clarified if differential pricing is to be facilitated.

5.1.3 Are global licensing agreements appropriate?

Finally, the geographical scope of any FRAND contract is also generally determined in light of 'sector practice' and may limit the prospect of pricing according to market conditions. As evidenced in UK litigation arising from FRAND disputes in the mobile communications sector, the commitment to license on FRAND terms, based on sector practice, can be a commitment to license on a global basis.[85] FRAND is, itself, a global commitment and thus when applied to multinational licensees with global sales the logical result is a worldwide licence. According to the Supreme Court, the cost and time involved in negotiating licences territory by territory is prohibitive and may act as a barrier to implementation of the standard. In addition, many consumers and businesses move with their devices across borders. Thus, given that implementers need to be sure that the rights acquired cover the use made of their devices by manufacturing partners and consumers, a global licence may be FRAND whereas a territory-by-territory approach may not.[86]

It is sensible, however, to question whether global licences are the most appropriate for the IoMT context when cost-effective access by consumers may actually require significant variation in SEP terms on a national or regional basis. It is true that, in many instances, global licensing may be more cost effective given the cumulative costs associated with licensing on a territory-by-territory basis.[87] Nevertheless, global licensing, theoretically at least, raises the spectre of a single royalty being set at a rate derived from comparable 'western' licenses. This concern was addressed by Birss J in the UK High Court decision in *Unwired Planet v Huawei*, where he observed that although '[a] worldwide licence would be far more efficient. It might well have different rates for different regions […] The employment of different rates would not lead the parties to abandon a worldwide license and go for country by country licensing'.[88] On the facts, Birss J held that a worldwide licence would be FRAND but have 'a different rate for sales in China' as 'comparable licences show that rates are often lower in China than for the rest of the world'.[89] Thus, where evidence exists of differential product pricing in

---

[85] *Unwired Planet v Huawei* (n 19) [15].
[86] ibid [15].
[87] *Vringo Infrastructure Inc v ZTE (UK) Ltd* [2015] EWHC 214 (Pat) [107].
[88] *Unwired Planet v Huawei* (n 38) [543].
[89] ibid [582]–[583].

different regions, this may be reflected in any FRAND licence, irrespective of its geographical scope. While FRAND licensing practice for the IoT/IoMT is still emerging, it is hoped that this same recognition of regional royalty variation will be similarly embraced outside of the mobile communications sector.

In summary, each of the above speaks to maintaining uniformity in licensing according to 'sector practice'. The vast majority of FRAND litigation has been confined to the mobile communication sector and principles developed according to practice in this sector. If the same principles are adopted for FRAND licensing between a licensor in the ICT sector and a licensee in the health technologies sector, this may create pricing and other barriers for an industry which has, in many instances, facilitated differential terms to provide wider access for a greater number of users. If manufacturers of IoMT devices/products are seen as 'similarly situated' entities to those in the mobile communications sector, and thus required to license SEPs on the same terms, the willingness of such firms to provide differential global pricing may be curtailed. This may hinder full-scale integration of the standard, a key objective of the FRAND commitment, and also stand in the way of widescale access to connected healthcare devices on a global scale.

5.2 FRAND and EU Competition Law

In contrast to contract law, competition law gives rise to both public and private enforcement mechanisms and, at first glance, may appear more suited to achieving the healthcare objectives set out above. Competition law scrutiny is not uncommon in the healthcare sector, with pharmaceutical companies regularly the subject of antitrust scrutiny by competition authorities and the courts.[90] In its 2019 report, the EU Commission emphasised the important role that competition law plays in controlling the price of medical treatments and in maintaining access to innovative healthcare treatments and medicines.[91] The report also acknowledged that competition authorities 'must remain vigilant and pro-active in investigating potentially anti-competitive situations, including where new practices used by companies or new trends in the industry are concerned'.[92] Accordingly, within the EU, pharmaceutical and related sectors

---

[90] In 2019, the EU Commission reported that between 2009 and 2017 the Commission and national competition authorities had adopted 29 antitrust decisions against pharmaceutical companies which imposed fines (totalling over 1 billion EUR) or made binding commitments to remedy anticompetitive behaviour. EU Commission, 'Competition Enforcement in the Pharmaceutical Sector (2009–2017)' COM (2019) 17 final.

[91] ibid.

[92] ibid 32.

can expect to be subject to continued competition law scrutiny. This must be welcomed, with the hope that its reach is wide enough, given the growing importance of the IoMT in providing effective medical treatment.[93] However, it is the actions of SEP holders that form the basis of this analysis and not the actions of IoMT producers.

Undoubtedly, simple analogies can clearly be drawn between the abusive licensing practices of certain pharmaceutical companies and SEP holders. The denial or the excessive cost of a licence, to either a generic pharmaceutical company or a standard implementer, blocks innovation by the licensee and denies the consumer access to the healthcare products they urgently need. Nevertheless, the extent to which 'public access' arguments may possibly influence the application of competition law in the FRAND context can only be assessed following a review of the law as currently applied to SEPs.

Under EU competition law, the refusal of an SEP holder to license on FRAND terms and/or to seek injunctive relief for patent infringement may constitute a breach of a dominant position under Article 102 TFEU. While the exercise of IPRs will only amount to abusive conduct in 'exceptional circumstances', the very nature of SEPs distinguishes them from other IPRs. According to the Court of Justice of the European Union in *Huawei v ZTE*, the *indispensability* of the SEP to competitors wishing to produce products that conform with the standard, together with the *irrevocable undertaking* made by the SEP holder to grant licences on FRAND terms in return for inclusion in the standard, creates 'particular circumstances'.[94] Crucially, these circumstances do not necessarily result in a finding of abuse. However, to avoid such a finding, they do require the SEP holder to respect the 'legitimate expectations' created by the FRAND commitment and to follow a framework for negotiation set out by the Court.[95] This procedural framework places obligations on the SEP holder to 'alert the alleged infringer of the infringement complained about' and present 'a specific, written offer for a licence on FRAND terms'.[96] It also creates obligations for the implementer to express

---

[93] For example, in July 2020, the EU Commission launched an inquiry into the sector of IoT for consumer products with a particular focus on the competitive landscape, developing trends and competition issues. EU Commission, 'Sector Inquiry into Consumer Internet of Things' C(2020) 4754 final. A preliminary report was released on 9 June 2021 and concluded that risks of 'gatekeepers' emerging controlling access to proprietary technology is a concern for the Commission, particularly in relation to voice assistant technology. EU Commission, 'Preliminary Report – Sector Inquiry into Consumer Internet of Things' SWD(2021) 144 final https://ec.europa.eu/commission/presscorner/detail/en/ip_21_2884 accessed August 2021.

[94] C-170/13 *Huawei v ZTE* (n 55) [48]–[52].

[95] ibid [53].

[96] ibid [61] & [63].

'its willingness to conclude a licensing agreement on FRAND terms' and to 'diligently [...] respond to the [SEP holder's] offer in accordance with recognised commercial practices in the field and in good faith'.[97] Any counter-offer must '[correspond] to FRAND terms' and, if rejected, the alleged infringer must provide 'appropriate security'.[98] In the event of continued disagreement 'the parties may, by common agreement, request that the amount of royalty be determined by an independent third party'.[99]

The procedural model established seeks to strike 'a fair balance between the interests concerned'.[100] However, while useful in assessing the abusiveness of parties' conduct and in defining a set of general guideposts for future FRAND negotiations, the framework has many shortcomings, which are further accentuated when considered in the context of the IoT and IoMT. First, the Court resists any attempt to determine FRAND-compliant terms or define their meaning. In doing so it implicitly rejects the argument that FRAND is a specific pricing commitment, leaving uncertain what a FRAND offer actually is and thus what is required to discharge the burden.[101] According to the Court, FRAND 'is a comity device that generates bilateral fair play obligations on both the patent owner and prospective licensees'.[102] It prefers to let the parties have recourse to an independent third party for adjudication of specific licensing terms, but acknowledges that this requires a willingness by both parties. Thus, expensive and time-consuming litigation may continue to act as a hurdle in solving FRAND disputes and a hurdle to market entry for innovative solutions, including healthcare treatments.

Second, the reasoning also appears limited to the specific context presented by the *Huawei v ZTE* dispute, that is, where parties are direct competitors in the marketplace.[103] As observed by Gallego and Drexl, the harm to competition that may arise from Huawei's request for an injunction is the exclusion of competitors from manufacturing products that compete with the claimant.[104] Yet, 'exclusionary abuse' – the leveraging of patents to bar direct competition – is not the harm arising in the IoT/IoMT setting.[105] In this context, the SEP

---

[97] ibid [63] & [65].
[98] ibid [66] & [67].
[99] ibid [68].
[100] ibid [55].
[101] Nicolas Petit, 'EU Competition Law Analysis of FRAND Disputes' in Jorge L Contreras (ed), *The Cambridge Handbook of Technical Standardization Law* (CUP 2017) 300.
[102] ibid.
[103] ibid 301.
[104] Gallego & Drexl (n 27) 147.
[105] Art 102(b) TFEU prohibits conduct which 'limit[s] production, markets or technical development to the prejudice of consumers'.

holder does not directly compete with firms in these markets.[106] In fact, this abuse can more correctly be characterised as 'exploitative abuse', that is, the leveraging of patents to prevent the production of something new.[107] In these circumstances the abusive conduct not only harms competition in the sector; it also harms innovation and thus denies consumers access to new (IoMT) products.[108] Coherency and consistency of practice demands that the relevant harm is identified and appropriately utilised as the measure against which abusive conduct is measured. Without this clarity, competition law will fail to prevent 'welfare-reducing' conduct, especially in emerging areas of contention such as the IoT.[109]

Finally, the potential for abusive conduct on both sides of a FRAND dispute, that is, SEP 'hold-up' and implementer 'hold-out', was influential in the reasoning of the CJEU in *Huawei v ZTE*. According to the Court, when applying Article 102 in the FRAND context, the implementer's conduct and that of the SEP holder are to be equally subject to scrutiny. Under what has been termed the 'willing licensee' test, implementers must demonstrate not only that the SEP holder has not fulfilled its FRAND obligations, but also that they have participated in the negotiation process with genuine intent. According to two recent interpretations by the German Federal Court of Justice in *Sisvel v Haier (I & II)*, a willing licensee must (i) clearly and unequivocally declare its unconditional readiness to accept a licence on FRAND terms, and (ii) purposefully participate in the negotiation process in a target-oriented manner to avoid delay.[110] The relevance of the implementer's conduct for the assessment gives a clear indication that a compulsory right of access to a SEP is not possible under competition law, even where the SEP holder has not acted in accordance with their FRAND commitment.[111] It also acknowledges that the SEP situation is different from earlier IPR/antitrust case law and that the potential for misconduct on both sides must be expressly factored into the analysis. The call for a 'fair balance' between the parties undermines the impact antitrust law has historically had on abusive use of the patent system and manifests itself in a weakening of competition law oversight. As observed

---

[106] Gallego & Drexl (n 27) 147.

[107] Article 102(a) TFEU prohibits conduct which 'directly or indirectly impos[es] unfair purchase or selling prices or other unfair trading conditions'.

[108] Giuseppe Colangelo & Robert Pardolesi, 'Intellectual Property, Standards, and Antitrust: A New Life for the Essential Facilities Doctrine? Some Insights from the Chinese Regulation' in Gustavo Ghidini, Hanns Ullrich & Peter Drahos, *Kritika: Essays on Intellectual Property* (Edward Elgar Publishing 2017) 70, 77.

[109] Gallego & Drexl (n 27) 151.

[110] Bundesgerichtshof [BGH] [Federal Court of Justice] Case No. KZR 36/17 5 May 2020 & Case No. KZR 35/17 24 November 2020 [57].

[111] Ullrich (n 56) [40].

by Rupprecht Podszun, '[t]he trump card of antitrust law (enshrined primarily in the TFEU), which is asserting itself against excesses in patent law, can no longer be played out as such. It needs "a fair balance". This makes it more difficult to intervene.'[112]

Thus, not only can competition law not give rise to a duty to license SEPs as a principle but the CJEU has also failed to reinforce its role in preventing abuse of IPRs, especially in the SEP context. In fact, the impression given to many is of a court 'that does not want to state too much and does not want to interfere more than necessary in an ongoing multinational business feud'.[113]

In light of the above it is possible to conclude that competition law principles, as currently applied in SEP/FRAND disputes, show little space for recognition of the wider policy issues at play. In fact, they appear ill suited to accommodate the connectivity issues likely to arise in the IoT context more generally. Without any direct engagement with the definition of FRAND and a clear understanding of the relevant harm that the law is designed to prevent, competition law may be of limited relevance in directly securing the public health objectives that, via the IoMT, find their way into the SEP/FRAND debate. Gallego and Drexl argue that competition law solutions can only arise in this context if the relevant harm is properly understood. They advocate for an 'effects-based' theory of harm where anticompetitive behaviour of the SEP holder is analysed in light of its impact on 'dynamic competition and innovation' rather than on whether or not they are competing in the secondary market. In this scenario the SEP's conduct, in seeking excessive royalties, is assessed in light of its impact on follow-on innovation by the implementer. It also evaluates the extent to which the SEP holder's conduct undermines trust in the standard-setting process, by reducing uptake and thus royalties for other SEP holders linked to that standard, thereby obstructing their incentive to innovate for the benefit of future standards.[114] Such an approach would, at least, require the court or authorities to focus on the need to protect innovation for downstream producers and consumers, and realign the analysis with the appropriate legal base. However, the practical aspects of securing access, that is, the relevant FRAND terms, will, for now, have to seek a resolution outside of the antitrust framework.

---

112 Podszun (n 27) 729.

113 Björn Lundqvist, 'The Interface between EU Competition Law and Standard Essential Patents: From *Orange Book Standard* to the *Huawei* case' (2015) 11 European Competition Journal 367, 385.

114 Gallego & Drexl (n 27) 149.

6. CONCLUSION

The IoMT is giving rise to a technical revolution in healthcare provision that promises to have a global impact. Yet, standard essential patents may obstruct that revolution and further reinforce the discord that exists between patents and access to healthcare treatments. Affordable licensing of formal technical standards will help to alleviate this tension and create one less hurdle to the global dissemination of valuable healthcare devices. The solution rests within the meaning of FRAND, and the above analysis has sought to investigate what room, if any, exists for consideration of the public health objectives within this private law obligation.

FRAND, as a contractual obligation, creates an interface between public and private governance. However, contract law by its very nature is limited to the interests directly embodied within the agreement. Nevertheless, the incompleteness of the FRAND obligation leaves room for wider public health concerns to be given effect. Practical considerations such as licensing level and geographical scope offer space for flexible solutions that can accommodate sector-specific concerns. In addition, the contemporary understanding of the non-discrimination limb supports sectoral variation in licensing terms. In contrast, competition law appears ill suited to take on this public governance role. Unwillingness by the CJEU to adopt a more interventionist approach currently leaves EU competition law at the margins. Without more considered engagement with the relevant harm arising in the FRAND/IoT/IoMT context, an effective solution from this legal base is unlikely to be forthcoming.

Consequently, current legal frameworks appear to offer some indirect solutions to the public health concerns raised by the IoMT. However, as demonstrated above, these are far from certain. Outside of the automotive sector, FRAND/IoT litigation has yet to surface. Yet, these same issues are a reality in other IoT sectors, giving rise to increased transaction costs and delaying the emergence of new technologies. Given that FRAND is a contractual instrument with an increasing role to play in fulfilling public policy objectives, greater regulatory oversight is required if formal standards are to facilitate, rather than hinder, the welfare benefits promised by the IoMT.

8. Accounting for ethical considerations in the licensing of patented biotechnologies and health-related technologies: a justification

Aisling McMahon[1]

1. INTRODUCTION

The role of 'ethics' in patent licensing could be used to refer to multiple aspects of how patented technologies are licensed in an 'ethical' manner (broadly defined),[2] under which two main categories are identified here and form the focus of this chapter, namely: (a) the licensing of patents in a manner which seeks to restrict ethically contentious *uses of underlying technologies subject to patent* – this form of licensing clause is often referred to as an 'ethical licensing' clause; and (b) the licensing of patents in a manner which seeks to encourage or facilitate ethical approaches to how *the patent right itself is used, that is, enforced or licensed*. More specifically, in this latter context, a key ethical issue that can arise is that patents (and other intellectual property rights) can be used in a way that impedes access to a patented technology. Accordingly, the chapter focuses in this context on patent licensing approaches whereby licensing conditions are adopted that seek to ameliorate

[1] An earlier version of the ethical licensing discussion in this chapter was presented at the ATRIP Conference, Vanderbilt University, Nashville, Tennessee, 26 August 2019; Durham CELLS Research Seminar, Durham University, 30 October 2019; Patents and Emerging Technologies Workshop, Wellcome Trust, London, 19–20 November 2019 and the Health Law and Ethics seminar series, Queen's University Belfast, 27 February 2020. I would like to thank the attendees for their helpful comments on these earlier drafts. This work on ethical licensing was generously supported by an Irish Research Council, New Foundations Grant 2018 for the project 'Patents and Biotechnology: Licensing as Ethical Governance?'

[2] The definition of what is 'ethical' in this context will depend on how those designing such clauses use or envisage such uses, which can give rise to controversy, as examined further below.

potential access issues related to patented health technologies (for example, by adopting licensing conditions which offer favourable terms to groups who may otherwise be unable to afford access, such as for low-income countries). For the purposes of clarity, the chapter will use the term 'ethical access licensing clauses' to describe such licensing clauses.

Where ethical considerations within patent licensing are discussed in the existing literature, the focus is often on how ethical issues could be considered in the licensing of specific patented technologies. For example, in the context of category (a) the literature on ethical licensing clauses has focused primarily on the uses of ethical licensing clauses in the context of genome editing technologies,[3] while in the context of category (b) the literature often focuses on access issues arising in relation to specific technologies, such as the impact of human gene patents on access to and delivery of genetic testing.[4] This chapter differs from such existing work by analysing the general justification for greater scrutiny of ethical issues in patent licensing. It builds on the author's existing work that argues that patents enable rightsholders to exercise a *governance* function over underlying technologies subject to patent,[5] which in turn enables rightsholders to become gatekeepers over the patented technology for the patent term. During this time, rightsholders can dictate how technologies under patent are used, and by whom, and can also control the terms of access to such technologies. Due to this gatekeeping role, this chapter argues that there

---

[3] For example, see: Christi J Guerrini, Margaret A Curnutte, Jacob S Sherkow and Christopher T Scott, 'The Rise of the Ethical License' (2017) 35(1) Nature 22; Jacob S. Sherkow, 'Patent Protection for CRISPR: An ELSI Review' (2017) 4(3) Journal of Law and the Biosciences 565; Aisling McMahon, 'Biotechnology, Health and Patents as Private Governance Tools: The Good, the Bad and the Potential for Ugly?' (2020) 3 Intellectual Property Quarterly 161; Oliver Feeney, Julian Cockbain, Michael Morrison, Lisa Diependaele, Kristof Van Assche and Sigrid Sterckx, 'Patenting Foundational Technologies: Lessons From CRISPR and Other Core Biotechnologies' (2018) 18(12) American Journal of Bioethics 36.

[4] S Aymé, G Matthijs, S Soini, et al, 'Patenting and Licensing in Genetic Testing' (2008) 16 European Journal of Human Genetics 405; Secretary's Advisory Committee on Genetics Health and Society, Gene Patents and Licensing Practices and Their Impact on Patient Access to Genetic Tests, 2010; Gertuii van Overwalle (ed.), *Gene Patents and Collaborative Licensing Models: Patent Pools, Clearinghouses, Open Source Models, and Liability Regimes* (Cambridge University Press 2009); Naomi Hawkins, 'The Impact of Human Gene Patents on Genetic Testing in the United Kingdom' (2011) 13(4) Genetic Medicine 320.

[5] Aisling McMahon, 'Biotechnology, Health and Patents as Private Governance Tools: The Good, the Bad and the Potential for Ugly?' (2020) 3 Intellectual Property Quarterly 161; Aisling McMahon, 'Global Equitable Access to Vaccines, Treatments and Diagnostics for Covid-19: The Role of Patents as Private Governance' (2021) 47(3) Journal of Medical Ethics 142.

are strong justifications for a reconsideration of: (a) the usefulness of patent licensing conditions as part of a broader governance or regulatory approach to address ethical issues posed by contentious uses of patented technologies (via ethical licensing clauses);[6] and (b) the need for greater use of patent licensing conditions which take into account ethical issues – including access issues, the core focus in this context of this chapter – related to how patents over such technologies are used to provide oversight over rightsholders' power where needed (via ethical access licensing clauses).

The chapter focuses specifically on patented biotechnologies and health technologies related to the human body as a case study. This case study is chosen because, as will be argued below, the rationale for considering ethical issues in the licensing of patents is heightened for such technologies due to the potential impact of patents' gatekeeping function for human health and well-being in this context.[7]

In making such arguments, the chapter is structured as follows. Section 2 builds the general case for why ethical considerations should be accounted for in the licensing of patented technologies related to the human body due to the gatekeeping function which patents place the rightsholder(s) in, and the potential impacts for human health. Following this, sections 3 and 4, respectively, focus on the two distinct categories of how ethical considerations could be considered at patent licensing stage, as indicated above, namely: category (a) to address ethical issues posed by *contentious uses* of patented technologies – in this context, ethical licensing conditions could be used as a complementary regulatory tool for emerging technologies; and category (b) to address ethical considerations arising around how rightsholder(s) *use patent rights* and their gatekeeping role over patented technologies to facilitate ethical *access to patented technologies*. Sections 3 and 4 build the specific case for why ethical considerations should be accounted for under both categories, and maps some

---

6 In terms of the meaning of 'governance', the chapter draws on Braithwaite et al's conception of governance as a broad category involving providing, distributing and regulating, with regulation being conceived of a 'large subset of governance that is about steering the flow of events and behaviour': see John Braithwaite, Cary Coglianese and David Levi-Faur, 'Can Regulation and Governance Make a Difference? [2007] Regulation and Governance 1.

7 In making such arguments, the chapter is not suggesting that ethical considerations are not impacted by the licensing of patented technologies which do not relate to the body. Indeed, the arguments made have resonance for a myriad of other contexts, for example in the context of access to patented carbon-reducing technologies, given that access to such technologies could be considered in the interests of the public. For a discussion of the role of intellectual property rights in the climate context, see: Abbe Brown, *Intellectual Property, Climate Change and Technology: Managing National Legal Intersections, Relationships and Conflicts* (Edward Elgar Publishing 2021).

of the key issues which need further investigation to ensure ethical considerations are embedded within patent licensing in a manner which aligns with the broader public interests at stake. Section 5 concludes, arguing that much deeper scrutiny and an interdisciplinary conversation is needed around the extent of rightsholders' control over patented technologies and how licensing conditions could be imposed to address potential ethical issues arising in this context. This should include greater scrutiny of how ethical licensing conditions could be used to complement existing regulatory approaches for steering ethical uses of patented technologies. It should also include greater scrutiny of the need for greater consideration (and normalisation) of ethical access licensing conditions being used on a more routine basis in other contexts to ensure access to downstream patented technologies is delivered in an ethical manner, if society is to realise the benefits of emerging technologies.

2. PATENT LICENSING AND TECHNOLOGIES RELATED TO THE HUMAN BODY: A JUSTIFICATION FOR GREATER SCRUTINY OF ETHICAL ISSUES

Three key interlinked arguments are put forward in this section to build the general case for why ethical issues should be given greater consideration and account within patent licensing for technologies related to the human body.

First, this section will argue that, once granted, a patent enables rightsholders to exercise a governance function over the technology under patent[8] – and that this function places rightsholders in a gatekeeping role over *uses* of the patented technology for the duration of the patent.[9]

Second, it argues that the impact of the gatekeeping role that rightsholders play over how technologies under patent are accessed and used has significant potential implications for human health and life, where the patented technology relates to the human body (for example, a patented medicine). Given such implications, it argues that how rightsholders exercise this governance function warrants scrutiny, including greater consideration of ethical issues at the patent licensing stage.

Third, and relatedly, it will argue that taking such ethical considerations into account at the licensing stage aligns with the broader social bargain and public interest rationale which many see as underpinning patent law. This is

---

[8] This is discussed further in: Aisling McMahon, 'Biotechnology, Health and Patents as Private Governance Tools: The Good, the Bad and the Potential for Ugly?' (2020) 3 Intellectual Property Quarterly 161.

[9] This is generally 20 years – see Article 33, TRIPS Agreement.

particularly true where *uses* of the underlying technology subject to patent raise contentious ethical questions and have the potential to pose harms to society. It is also the case for patented health technologies where certain uses of the *patent right* (as opposed to the technology under patent) could hinder or stifle downstream access to the technology, thereby affecting human health.

2.1 The Governance Function of Patents: Rightsholders as Gatekeepers

A patent gives the rightsholder a right to exclude others from using a patented technology for the duration of the patent – generally 20 years.[10] This exclusionary right granted by a patent is often discussed in terms of its economic function and as providing an incentive for the development of an invention.[11] This is because once patented, third parties must generally seek permission (via a contractual licence) from the rightsholder to use a patented technology. This exclusivity right allows the rightsholder to derive an income stream from the patent. If a third party uses a technology without the rightsholder's permission, barring limited exceptions applying,[12] that third party could be liable for patent

---

[10] The protection conferred by a patent can be extended in some contexts; for example, in the European context, a supplementary protection certificate (SPC) can be applied for medicinal and some plant protection products. Where granted, this extends the patent protection period by up to five years. See generally: www.ipoi.gov.ie/en/types-of-ip/supplementary-protection-certificates/spc-s-explained/ accessed July 2021.

[11] See discussion of patents and innovation in: Edwin Mansfield, 'Patents and Innovation: An Empirical Study' (1986) 32(2) Management Science 173; The role of patents as incentives is subject to critique – see: James Bessen and Michael J Meurer, *Patent Failure: How Judges, Bureaucrats, and Lawyers Put Innovators at Risk* (Princeton University Press 2009). For a discussion of patents' role in incentivising innovation and differences across technology sectors, see: Mark A Lemley and Dan L Burk, 'Policy Levers in Patent Law' (2003) 89 Virginia Law Review 1575, Minnesota Public Law Research Paper No. 03-11, UC Berkeley Public Law Research Paper No. 135, http://dx.doi.org/10.2139/ssrn.431360. See also discussion in: Siva Thambisetty, Aisling McMahon, Luke McDonagh, Hyo Yoon Kang and Graham Dutfield, 'The TRIPS Intellectual Property Waiver Proposal: Creating the Right Incentives in Patent Law and Politics to End the COVID-19 Pandemic' (24 May 2021) LSE Legal Studies Working Paper No. 06/2021.

[12] For example, a research exemption could apply. Such an exemption allows use of a technology for research purposes. It should be noted that the scope of such exemptions varies by jurisdiction. In the UK context, see s 60(5)(b) of the Patents Act 1977, as amended. In the UK this exemption is interpreted narrowly, wherein any uses for research purposes which could lead to a commercial use would generally not be covered by it. See: Aisling McMahon, 'Patents, Human Biobanks and Access to Health Benefits: Bridging the Public–Private Divide' in Jessica Lai and Antoinette Maget Dominicé (eds) *Intellectual Property and Access to Im/Material Goods* (Edward Elgar

infringement. However, once a patent is granted it also gives rightsholders a significant role in dictating how that technology under patent is used, and by whom, for the duration of the patent. As I have argued elsewhere,[13] this gives rightsholders a private governance role over the technology, because how the patent is licensed (and enforced) by the rightsholder(s), if at all,[14] will impact who can access the patented technology downstream and on what terms. Rightsholders' decisions over how the patent is enforced and licensed also impact how the technology under patent is developed downstream (during the patent term). Moreover, if the technology under patent is required for other technologies to operate – for example, if the underlying technology is a platform technology or a process – access to that technology may be needed for multiple other technologies to operate. Thus, how the rightsholder(s) uses the patent right, and how they choose to license that right (if they choose to license it to third parties at all), can significantly impact how other technologies are used and developed during the patent term.[15]

The exclusivity which underpins the patent right places the rightsholder(s) in a role that is akin to a *gatekeeper of the technology under patent* for the patent term – because due to the patent right they can control or gatekeep who can access and use the patented technology and on what terms.[16] Rightsholders can also limit uses of the patented technology – for example, by imposing licensing clauses prohibiting certain uses of a patented technology within patent

---

Publishing 2016). In the human genome editing context, the role of research exemptions is discussed in: Duncan Matthews, Abbe Brown, Emanuela Gambini, Aisling McMahon, Timo Minssen, Ana Nordberg, Jacob S. Sherkow, Jakob Wested, and Esther van Zimmeren, 'The Role of Patents and Licensing in the Governance of Human Genome Editing: A White Paper' (30 July 2021) Queen Mary Law Research Paper No. 364/2021.

[13] Aisling McMahon, 'Biotechnology, Health and Patents as Private Governance Tools: The Good, the Bad and the Potential for Ugly?' (2020) 3 Intellectual Property Quarterly 161.

[14] A rightsholder may also refuse to license a patented technology to third parties.

[15] Aisling McMahon, 'Biotechnology, Health and Patents as Private Governance Tools: The Good, the Bad and the Potential for Ugly?' (2020) 3 Intellectual Property Quarterly 161.

[16] Notably, the patent right does not confer the rightsholder with a right to use an invention per se, as a patent is a negative right to stop others from using the technology, but a patent grant does not necessarily mean the rightsholder can use that technology under patent – see: Aisling McMahon, 'Biotechnology, Health and Patents as Private Governance Tools: The Good, the Bad and the Potential for Ugly?' (2020) 3 Intellectual Property Quarterly 161. Other regulations may affect how a technology can be used or worked; for instance, national regulatory frameworks may prohibit certain uses of an emerging technology. Thus, even if a rightsholder does not restrict a certain use in a licensing clause, this may be prohibited by national or international regulation.

licences. It is acknowledged that a rightsholder's ability to impose such clauses may depend on the respective licensing parties' freedom of contract, as in this context the strength of the licensee and licensor's relative bargaining positions will be key (a point discussed further in section 3 below).[17] Nonetheless, the potential for rightsholders to introduce such clauses at licensing stage for a patented technology is always latent. This gatekeeping role enabled by the patent right is the starting point for justifying why ethical considerations should be given greater at patent licensing stage.

Moreover, this gatekeeping function is double-edged in nature, and as a result it could be a useful feature for a broader regulatory approach to steer health technologies but also a potentially harmful feature in hindering access to emerging health technologies. Both aspects of this function enabled by the patent right require greater scrutiny in order to embed ethical issues within patent licensing.

In this context, on the one hand, the gatekeeping function provided by the patent could be harnessed and used within a broader regulatory context to restrict ethically contentious uses of a technology under patent. In other words, it could be channelled in a positive way to limit or restrict ethically contentious *uses of patented technologies*, where warranted (via ethical licensing clauses). This would provide an avenue for patents to be used in a manner that complements existing regulatory approaches for emerging technologies (discussed in section 3).

On the other hand, rightsholders' gatekeeping role can have significant (potentially negative) impacts on access to patented technologies; this aspect also requires greater scrutiny. In this context, as will be discussed below, *how a rightsholder licenses (or may refuse to license) a patented technology* can significantly impact, for example, *access to healthcare* where technologies under patent relate to the human body. In such contexts, greater scrutiny over the rightsholder's ability to licence a technology is warranted, including

---

[17] This chapter also acknowledges that other forms of intellectual property rights are relevant in how technologies are accessed and used, particularly in the health context, where trade secret protection is often a key factor in relation to access to technologies – for example, for vaccines. For a discussion of this in the COVID-19 context, see: Siva Thambisetty, Aisling McMahon, Luke McDonagh, Hyo Yoon Kang and Graham Dutfield, 'The TRIPS Intellectual Property Waiver Proposal: Creating the Right Incentives in Patent Law and Politics to End the COVID-19 Pandemic' (24 May 2021) LSE Legal Studies Working Paper No. 06/2021; Karen Walsh, Andrea Wallace, Mathilde Pavis, Natalie Olszowy, James Griffin and Naomi Hawkins, 'Intellectual Property Rights and Access in Crisis' (2021) 52 IIC 379; Allison Durkin, Patricia Anne Sta Maria, Brandon Willmore, and Amy Kapczynski, 'Addressing the Risks That Trade Secret Protections Pose for Health and Rights' (2021) 23(1) Health and Human Rights Journal 129.

consideration of how to embed ethical approaches to licensing which facilitate broader access to patented technologies particularly in the health context (via ethical access licensing clauses).

2.2 Licensing Patented Technologies and the Patent Social Bargain: Technologies Related to the Human Body

Building on the foregoing, this section argues that the gatekeeping role that rightsholders occupy due to the patent right warrants particular scrutiny in the context of patented technologies related to the human body for both categories indicated above, with the general justification for this considered separately for each category here, namely: (i) due to the significant implications of how the patent or the underlying technology, respectively, is used for human health and life; and (ii) due to the social bargain justification underpinning patent law.

First, if we consider category (b) and the ethical issues related to access to patented technologies, as noted above, rightsholders' gatekeeping role means their decisions around how they use a *patent right* can significantly impact access to patented technologies. Where the patented technology is a health-related technology, such decisions can hinder access to healthcare. The impacts of how rightsholders use patents (and other intellectual property rights) in the healthcare context are well-documented and can be illustrated by numerous examples. For instance, the potential impact of intellectual property rights on healthcare is starkly illustrated by the HIV/AIDS crisis in the 1990s and early 2000s and the ensuing discussions around the adoption of the Doha Agreement in 2001.[18] At the time, anti-retroviral medicines which could be used to treat HIV/AIDS were patented. Thus, rightsholders had significant control over the terms of access to such medicines, and because such patented medicines were in many cases priced far in excess of what many people affected by HIV/AIDS could afford, such treatments were inaccessible for millions particularly in Africa, with fatal consequences.[19] Moreover, at the time of writing, international debates continue around the role of patents (and other intellectual property rights) for access to COVID-19 vaccines and other health technologies. In this context, a significant gap has emerged around

[18] See discussion in Ellen 't Hoen, Jonathan Berger, Alexandra Calmy and Suerie Moon, 'Driving a Decade of Change: HIV/AIDS, Patents and Access to Medicines for All' (2011) 14(1) Journal of the International AIDS Society 14.

[19] Aisling McMahon and Edana Richardson, 'Patents, Healthcare & Engaged Shareholders: A Pathway to Encourage Socially Responsible Patent Use?' (2021) Legal Studies (FirstView) 1. doi:10.1017/lst.2021.49.

access to COVID-19 vaccines between high-income countries (HICs)[20] and low and middle-income countries (LMICs) – many LMICs will not have sufficient access to vaccines for their population until 2023 or later, based on recent predictions.[21] Although patents and other intellectual property rights are not the only roadblock to access to COVID-19 vaccines and other health technologies,[22] they nonetheless play a significant role. Moreover, within the pandemic context, rightsholders have acted as gatekeepers exercising significant control over which State will gain access to COVID-19 vaccines first and on what terms.[23] This governance function which rightsholders hold in this context has had particularly significant implications due to the shortage of vaccines to meet global population needs, and the demand for vaccines to control COVID-19 everywhere.

Many other examples of the impact of rightsholders' decisions around how patents are used or licensed for access to healthcare could be given, in both health emergency and everyday healthcare contexts. For instance, Chapters 2 and 3 of this volume examine the role of patents for access to and delivery of non-invasive prenatal testing (NIPT).[24] In short, such examples illustrate that how patents over such health technologies are used can significantly affect

---

[20] This is correct at the time of writing in September 2021. In this context, the discussion in many high-income countries is now around the use of booster doses of vaccines, while many low and middle income countries have not been able to access first doses of vaccines for their populations. The World Health Organization initially called for a moratorium on vaccine booster doses until the end of 2021: 'WHO Director-General's opening remarks at the media briefing on COVID-19 – 8 September 2021' available at www.who.int/director-general/speeches/detail/who-director-general-s-opening-remarks-at-the-media-briefing-on-covid-19---8-september-2021 accessed 9 September 2021; Nature Editorial, 'The WHO Is Right to Call a Temporary Halt to COVID Vaccine Boosters' (2021) 596 Nature 317.

[21] Oxfam, 'Campaigners Warn that 9 out of 10 People in Poor Countries Are Set to Miss Out on COVID-19 Vaccine Next Year' (9 December 2020) available at https://www.oxfam.org/en/press-releases/campaigners-warn-9-out-10-people-poor-countries-are-set-miss-out-covid-19-vaccine (accessed 9 September 2021).

[22] See: Siva Thambisetty, Aisling McMahon, Luke McDonagh, Hyo Yoon Kang and Graham Dutfield (2021) The TRIPS Intellectual Property Waiver Proposal: Creating the Right Incentives in Patent Law and Politics to end the COVID-19 Pandemic' (24 May 2021) LSE Legal Studies Working Paper No. 06/2021.

[23] Aisling McMahon, 'Global Equitable Access to Vaccines, Medicines and Diagnostics for COVID-19: The Role of Patents as Private Governance' (2021) 47 Journal of Medical Ethics 142.

[24] Dianne Nicol and Jane Nielsen, 'Non-invasive Prenatal Testing and the Resilience of the Patent System' in Naomi Hawkins (ed.) *Patenting Biotechnical Innovation: Eligibility, Ethics and Public Interest* (Edward Elgar Publishing 2022) chapter 3; Andrew Sagar, Kathleen Liddell, Johnathon Liddicoat and Mateo Aboy, 'Patent Eligibility and Precision Medicine: Emerging Technologies, Legal Flux and

access to healthcare. This chapter argues that such impacts are a key reason why the ethical implications of how patents over such technologies are used, and specifically how this affects access to health technologies should be given greater consideration at patent licensing stage. [25]

Relatedly, a further rationale for greater scrutiny over rightsholders' licensing practices and for greater focus on embedding ethical considerations around access to healthcare within patent licensing is evident if we consider the underlying social bargain theory of patents.[26] Under the patent bargain rationale, the grant of a patent is conceived of not as an automatic right but rather as a limited exclusive right given to the rightsholder in return for disclosure of knowledge that is of benefit to society.[27] The patent social bargain refers to this *quid pro quo* underpinning patent grant. Patents, unlike other forms of intellectual property rights such as copyright, are not granted automatically. Instead, the grant is subject to the patent application meeting patentability criteria of inventive step, novelty and technical application; only then will a rightsholder be granted a patent over a technology. That patent grant also comes with conditions requiring the disclosure of information around how to work the technology in the patent application. It is conceded on this point that the effective operation of multiple aspects of the patent social bargain could be called into question

---

Empirical Methods' in Naomi Hawkins (ed.) *Patenting Biotechnical Innovation: Eligibility, Ethics and Public Interest* (Edward Elgar Publishing 2022) chapter 2.

[25] This argument is not to suggest that in some cases denial of the patent under Art 53(a) EPC will not be needed.

[26] The patent bargain is outlined by: William Robinson, *The Law of Patents for Useful Inventions* (Boston: Little, Brown, and Co., 1890), Vol. II, at 70 as follows: 'A patent is a contract between the inventor and the public, by which the inventor, in consideration that the exclusive use of his invention is secured to him for a limited period of time, confers upon the public the knowledge of the invention during that period and an unrestricted right to use it after that period has expired [...] The specification is the instrument in which the terms of these mutual considerations and promises are declared, and on its completeness and accuracy depends the validity and the value of the contract itself.' See generally, Mario Biagioli, 'Weighing Intellectual Property: Can We Balance the Social Costs and Benefits of Patenting?' (2019) 57(1) History of Science 140. The social bargain underpinning patents is discussed in Karen Walsh, 'Invoking the Public Interest in the European Patent System' in Naomi Hawkins (ed.) *Patenting Biotechnical Innovation: Eligibility, Ethics and Public Interest* (Edward Elgar Publishing 2022); see also 'Introduction' in Naomi Hawkins (ed.) *Patenting Biotechnical Innovation: Eligibility, Ethics and Public Interest* (Edward Elgar Publishing 2022).

[27] See discussion in: Margaret Llewelyn, 'Schrodinger's Cat: An Observation on Modern Patent Law' in Peter Drahos *Death of Patents* (Lawtext Publishing 2005). For a critique of the patent bargain theory, see Shubha Ghosh, 'Patents and the Regulatory State: Rethinking the Patent Bargain Metaphor after Eldred' (2004) 19(4) Berkeley Technology Law Journal 1315.

within contemporary patent practice, including the extent to which patent disclosure requirements provide sufficient information to facilitate the working of technologies.[28] However, such issues are beyond the scope of this discussion; instead, here, the focus is on this ideal of a patent social bargain within patent law. If we view patents from the perspective of this approach, it arguably implies that the grant of a patent should in the general sense be in the interests of society.

If, however, a patent is granted over a health technology, and that patent is used by the rightsholder in a manner which disproportionately hinders access to healthcare, this gives rise to questions about how the patent social bargain is working. More specifically, if a patent right is being used in a manner that results in a technology arising which has significant health benefits (or which is essential for human life to be sustained) but which cannot be accessed by those who most need it, we must question how the social bargain underpinning patent law is working. In such contexts, this chapter argues that the social bargain theory provides an additional justification for why ethical considerations should be given greater oversight at patent licensing stage to ensure patents are licensed in a manner which does not disproportionately hinder access to patented technologies. These points are developed further in section 4.

We turn now to the second category, ethical licensing clauses and the general case for why ethical considerations should be accounted for *in how the technology under patent is used*. In this context, patents may be granted over technologies where certain uses of that technology pose significant ethical implications for human life or health. For instance, certain uses of an emerging technology could fundamentally change or disrupt society and/or the existing ecosystem, such as via the use of emerging biotechnologies to create novel beings that may threaten life for humans and other species[29] or the uses of emerging technologies, such as gene editing or cloning to raise possibilities of recreating former human species or beings which resemble them, giving rise to significant ethical implications.[30] Such implications provide a practical

---

[28] For a discussion of the limitations of patent disclosure in the COVID-19 vaccine context see: Siva Thambisetty, Aisling McMahon, Luke McDonagh, Hyo Yoon Kang and Graham Dutfield (2021) The TRIPS Intellectual Property Waiver Proposal: Creating the Right Incentives in Patent Law and Politics to End the COVID-19 Pandemic (24 May 2021) LSE Legal Studies Working Paper No. 06/2021.

[29] David Lawrence and Sarah Morley, 'Novel Beings: Moral Status and Regulation' (2021) 30(3) Cambridge Quarterly of Healthcare Ethics 415.

[30] Aisling McMahon, 'Patents, Governance and Control: Ethics and the Patentability of Novel Beings and Advanced Biotechnologies in Europe' (2021) 30(3) Cambridge Quarterly of Healthcare Ethics 529.

rationale for using ethical licensing clauses for patented technologies seek to discourage ethically contentious uses of such technologies as a complement to existing regulatory approaches, a point developed in section 3 below.

Moreover, in these and other contexts, if we again reflect on the social bargain model, this arguably provides further justification for why the patent licensing stage should be used to incorporate ethical clauses restricting contentious uses of such technologies. In such contexts, granting a patent if it could have a use which has the potential to pose significant harms for society – depending on the context and likely level of harm – could be seen as contrary to the patent bargain. If one were to apply the patent social bargain rationale to such contexts, either no patent should be granted – however, this would mean the other potentially beneficial uses of that technology would not be accounted for by the system – or a patent could be granted but clauses could be attached at licensing stage which restrict specific ethically contentious uses of that technology.

Furthermore, and relatedly, some would view the grant of a patent as a government (as patents are national rights) stamp of approval for an invention.[31] However, if certain uses of that technology could prove significantly harmful to society, it is questionable if the government should be giving a stamp of approval to such technologies without conditions attaching around how the patent is licensed - related to how the patented technology is used.[32] As will be discussed in section 3 below, in such contexts the adoption of ethical clauses within patent licences could provide a mechanism to allow the patent right in a general sense while discouraging or prohibiting contentious uses of the patented technology.

This section has argued that the potential implications of the gatekeeping role rightsholders play over patented technologies related to the human body (as enabled by the patent right), and its interaction with the patent social bargain, build a general case for why ethical considerations should be given greater scrutiny at the patent licensing stage. In sections 3 and 4 the chapter will build a more specific case for why there is a need for greater scrutiny of how ethical considerations are accounted for in the licensing of patents over such technologies, in the context of: category (a), licensing conditions to address the ethical implications posed by the *uses of the technology* under patent; and

---

[31] For a discussion, see: Catarina Holtz, 'Due Process for Industrial Property, European patenting under human rights control' (PhD Dissertation 2003) at 517–19.

[32] There may also be a need for conditions to be adopted around uses of the technology by other regulatory systems within a State; however, when a patent is being considered at grant stage, technologies are emerging and so such broader regulatory conditions on use of that technology may not be in place immediately or soon after the patent grant decision.

category (b), licensing conditions aimed at addressing ethical implications that rightsholders' *uses of the patent right* have for access to technologies, respectively. In both sections the discussion will highlight why refusal of a patent grant (in some cases) may not be the most appropriate avenue to address ethical issues posed (depending on the ethical issue at stake). In making such arguments, importantly, the chapter is not arguing that refusal of patent grant based on ethical issues arising such as under the morality or *ordre public* provisions in Europe should never occur – there will likely be cases where refusal of a patent based on ethical issues is wholly needed and warranted.[33]

3. PATENTS AND ETHICAL LICENSING CLAUSES RESTRICTING USES OF A PATENTED TECHNOLOGY: A JUSTIFICATION

Turning first to category (a) and how patent licensing clauses could - and more specifically arguably should be - be tailored to discouraging ethically contentious *uses of technologies* under patent.[34] This section will first provide a brief overview of how ethical licensing conditions can be used for such purposes (to date where such clauses are used this has primarily operated in a patent holder-driven manner). Following this, it will indicate concerns that arise in relation to patent holder-driven approaches to designing and adopting ethical licensing clauses. Finally, it will put forward a proposal around the potential for ethical licensing clauses to be used in a broader regulatory manner at a national or international level to complement other regulatory approaches to emerging technologies, and the key potential benefits from using patent licensing clauses in this manner. In doing so, this section's core aim is to further develop the case for, and role of, ethical licensing clauses as an avenue which could be used to discourage ethically contentious uses of patented technologies. It also calls for a deeper interdisciplinary conversation around the potential role of ethical licensing clauses across intellectual property, ethics and scientific communities.

---

[33] For a more detailed discussion of the morality provisions in European patent law, see: Aisling McMahon, 'The Morality Provisions in the European Patent System: An Institutional Examination' (PhD Thesis, University of Edinburgh, 2016).

[34] The term 'ethically contentious' is used here to refer to uses which give rise to ethical concerns, and particularly uses which pose safety concerns to human, animal life or the environment. This is a deliberately broad definition, as such clauses could be used for a myriad of purposes to ameliorate 'ethical' issues arising.

3.1 How Could Patent Licensing Clauses Be Used to Discourage Ethically Contentious Uses of Patented Technologies?

The nature of a patent licence is that it specifies the conditions under which a patented technology can be used by a third party via contractual terms. A patent licence is generally seen as an enabling tool, because patent licences enable or facilitate use of a technology subject to patent by third party licensees. Using a patented technology without a patent licence, as noted above, can render a party liable for patent infringement – unless a relevant exemption applies which allows use of that patented technology outside the permission conferred by the patent licence.[35] A patent infringement challenge could lead to costly litigation and/or the award of damages against a third party found to be infringing a patent.[36] However, given the gatekeeping role rightsholders play due to the patent right, patent licences can also be used in a restrictive manner to insert clauses into a patent licence which restrict *uses* of a technology under patent. As noted above, this could be achieved by the rightsholder inserting a condition within the patent licence which, for example, prohibits or excludes certain uses of a patented technology by the licensee(s). For instance, such clauses could be used by rightsholders if they were concerned about certain ethically contentious uses of a patented technology to prohibit such uses under the licence, with the aim of deterring or discouraging such uses. Patent licensing conditions which seek to restrict specific uses of a technology for such reasons are generally termed 'ethical licensing' clauses.

At a practical level, it is acknowledged that it may not be possible or practically feasible for an ethical licensing clause in the patent agreement to prohibit all ethically contentious uses of patented technologies, for several reasons. First, even if a patent is applicable in one jurisdiction, some entities or individuals may use the technology under patent by choosing to ignore the patent right.[37] Doing this would likely be in breach of the patent itself in that jurisdiction, and so could render that entity or individual liable for patent infringement, but penalties arising from this would depend on the infringement

---

[35] For example, a third party might use a technology under patent for research purposes and a research exemption may apply. See note 12 above.

[36] In the UK context, see: Patents Act 1977 as amended, s. 60 of which defines what constitutes patent infringement. Section 61 specifies remedies which can be applied for where patent infringement occurs, including: an injunction restraining further infringement, damages in respect of infringement, an account for profits derived by infringement.

[37] See generally Naomi Hawkins' work, which highlights that people or entities can sometimes act with 'wilful blindness' in relation to patents applicable in a field: Naomi Hawkins, 'The Impact of Human Gene Patents on Genetic Testing in the United Kingdom' (2011) 13 Genetic Medicine 320.

being identified and a legal action being taken against that entity. Some entities may choose to ignore a patent and run this risk of a patent infringement challenge. In such cases, an ethical licensing clause would not bind such an entity, as they would not have sought a patent licence in the first instance. However, this issue applies to all patented technologies and is not only pertinent in the ethical licensing context. Second, it is also similarly plausible that some people or entities who license a patented technology will be unwilling to abide by or choose to ignore the ethical licensing clause within that licence and take the risk of being faced with the penalty for breaching the licensing clause. Nonetheless, this is the same risk of non-compliance that one encounters with any regulatory proposal designed to shift a party's behaviour, and having a clause would at least provide a potential deterrent function for such uses. Third, patents are national rights, and patents may not apply over the technology in all jurisdictions. Where no patent is applicable in a region or country, no licence will be needed for use of that technology, and hence the ethical licensing clause will not have an effect in such jurisdictions.[38] Nonetheless, despite such potential limitations, ethical licensing clauses offer a useful means to discourage ethically contentious uses in some cases where a technology is patented, and as such could offer a complement to existing regulatory systems. Importantly, this chapter is not suggesting that such clauses should be seen as a replacement for other regulatory mechanisms around uses of emerging technologies; rather it argues that ethical licensing clauses in patent agreements offer a useful addition to existing frameworks, where they are used in a broader regulatory manner for the reasons given below.

A recent example of the use of ethical licensing clauses was the Broad Institute's policy over licensing its patents related to CRISPR-Cas9 technology for agricultural purposes.[39] In this context, the Broad Institute's licensing agreements related to agricultural uses of CRISPR-Cas9 technologies provided non-exclusive research and commercial licences related to such technologies, subject to three key restrictions, namely that the licences: prohibited uses of the licensed technology for gene drives; prohibited use of the licensed technology in a way which could create terminator or sterile seeds; and prohibited uses of

---

[38] See discussion of such jurisdictional differences for patents in the human genome editing context, in Duncan Matthews, Abbe Brown, Emanuela Gambini, Aisling McMahon, Timo Minssen, Ana Nordberg, Jacob S. Sherkow, Jakob Wested, and Esther van Zimmeren, 'The Role of Patents and Licensing in the Governance of Human Genome Editing: A White Paper' (30 July 2021) Queen Mary Law Research Paper No. 364/2021 available at https://ssrn.com/abstract=3896308 accessed 31 July 2021, 49.

[39] See discussion in: Aisling McMahon, 'Biotechnology, Health and Patents as Private Governance Tools: The Good, the Bad and the Potential for Ugly? (2020) 3 Intellectual Property Quarterly 161.

the licensed technology to modify tobacco with the exception of allowing such uses in the research context not directed to the commercialisation of tobacco, and in the context of manufacturing use for non-tobacco products.[40]

In reference to the use of such licensing clauses in this context, Issi Rozen, the Chief Business Operating Officer of the Broad Institute, noted that 'the use of genome editing in agriculture raises important ethical and safety concerns'.[41] Alongside this, he noted that such clauses were adopted following consultation with 'external experts and careful internal consideration' and that, while such issues were already within the oversight of US regulatory authorities, 'the Broad feels it is important to include explicit restrictions in the technology licenses as well'.[42] This example suggests that concerns about the ethical and safety implications of certain uses of the patented technology were a key motivator for adopting such restrictions.

3.2 Patent Holder-led Approaches to Ethical Licensing: Cause for Concern?

The Broad Institute example highlights the use of ethical licensing clauses within the patent context in a patent holder-led manner which appeared to be aimed at pre-empting or discouraging uses by third parties of a patented technology which the rightsholder, following consultation with others, viewed as raising ethical and/or safety concerns. This is a laudable approach. However, patent holder-led approaches to ethical licensing clauses (even where entirely well-intentioned) can in some cases raise ethical issues, including issues around the appropriateness of a rightsholder being the entity which decides on whether such clauses are adopted and what the content of such clauses should entail.

In this context, the clauses adopted in the CRISPR-Cas9 context by the Broad Institute were largely non-contentious, as the restrictions adopted were already issues which had generated controversies in the field and on which there was considerable agreement. However, it is plausible that a rightsholder could adopt conditions around technological use of a patented technology on which there is no consensus. Such conditions could for instance be adopted for a range of reasons, including, due to moral opposition to a particular technology by the rightsholder, and this could have a significant impact on the development and access to that technology. For instance, if certain uses of a novel

40 www.broadinstitute.org/news/licensing-crispr-agriculture-policy-considerations accessed July 2021.

41 Ibid.

42 Ibid.

technology were restricted by the rightsholder, this could have a blocking effect on other researchers or entities seeking to use that technology or develop it for commercial uses. If that technology was a health-related technology this could limit the potential access to and development of such health technologies for the duration of the patent.[43] Moreover, patent licence terms are often not published and can lack transparency,[44] and even where they are published, elements of such licenses may be redacted for commercial reasons. This in turn means that if ethical licensing clauses were used by rightsholders more frequently, the use and scope of such clauses could be shielded from broader scrutiny, leading to limited potential for external oversight over the purpose and effect of how such 'ethical' clauses are designed and the impact of their use. Relatedly, in this vein, there have also been concerns about the potential for a lack of democratic input in the design of ethical licensing clauses,[45] including concerns around the need for such clauses, where used, to be informed by key stakeholders and also by democratically elected representatives at a national or international level.

---

[43] See discussion in the context of abortion technologies, in: Kara W Swanson, 'Patents, Politics and Abortion' in Willliam T Gallagher and Debora J Halbert (eds) *Intellectual Property Law in Context: Law and Society Perspectives on Intellectual Property*. Northeastern University School of Law Research Paper No. 161-2013, available at https://ssrn.com/abstract=2337062 accessed 31 July 2021; Scott A Allen, 'Patents Fettering Reproductive Rights' (2012) 87(1) Indiana Law Journal 445; Mindy J Lees, 'I Want a New Drug: RU-486 and the Right to Choose' (1990) 63 Southern California Law Review 1113.

[44] In the Covid-19 context, see: Duncan Matthews, 'Coronavirus: How Countries Aim to Get the Vaccine First by Cutting Opaque Supply Deals' *The Conversation* (27 July 2020) available at https://theconversation.com/coronavirus-how-countries-aim-to-get-the-vaccine-first-by-cutting-opaque-supply-deals-143366 accessed 31 July 2021. See also: Médecins Sans Frontières Technical Briefing Document, Voluntary Licenses and Access to Medicines (October 2020) available at https://msfaccess.org/sites/default/files/2020-10/IP_VoluntaryLicenses_full-brief_Oct2020_ENG.pdf accessed 31 July 2021.

[45] WHO Expert Advisory Committee on Developing Global Standards for Governance and Oversight of Human Genome Editing, *Human Genome Editing a Framework for Governance* (July 2021) para 84; Duncan Matthews, Abbe Brown, Emanuela Gambini, Aisling McMahon, Timo Minssen, Ana Nordberg, Jacob S. Sherkow, Jakob Wested, and Esther van Zimmeren, 'The Role of Patents and Licensing in the Governance of Human Genome Editing: A White Paper' (30 July 2021) Queen Mary Law Research Paper No. 364/202, 48 available at https://ssrn.com/abstract=3896308 accessed 31 July 2021; see also: Naomi Scheinerman and Jacob S Sherkow, 'Governance Choices of Genome Editing Patents' (2021) *New Frontiers in Political Science* (6 September 2021) at 5; Oliver Feeney, Julian Cockbain and Sigrid Sterckx, 'Ethics, Patents and Genome Editing: A Critical Assessment of Three Options of Technology Governance' (2021) Frontiers in Political Science (21 September 2021) at 5.

Moreover, there are also questions around the likelihood that rightsholders would adopt ethical licensing clauses on their own initiative. In this context, as discussed elsewhere,[46] the Broad Institute is a peculiar entity – in comparison to many rightsholders in the biotechnology context – in the sense that it is a non-profit research organisation,[47] whose mission is to 'propel the understanding and treatment of human disease by tackling the most critical challenges in biology and medicine'.[48] This non-profit nature and its broader public health mandate may make it more likely that an institute of this nature would seek to adopt such a clause of its own initiative, and one which aligns with broader health interests, given its core mission. However, in contrast, many rightsholders, particularly in the biotechnology context, are likely to be for-profit entities. Furthermore, in the case of academic institutions, licensing may be assigned to surrogate licensing companies, which also tend to operate on a for-profit basis.[49] In such contexts it is questionable if such rightsholders would develop ethical licensing clauses on their own initiative without an external driver to encourage the adoption of such clauses. If they were to develop such clauses of their own accord, because such clauses generally restrict uses of a patented technology, this could potentially make the technology under patent less attractive as a technology to license for a third party. Hence, it could reduce the likely financial benefits which may derive from licensing such a technology.[50] This could give rise to tension or conflict in such contexts in the case of for-profit entities between the aim of maximising profits, and the potential impact on this of the use of an ethical licensing clause which would restrict certain uses of a technology. This issue again raises questions around the likelihood and appropriateness of rightsholder-led design of ethical licensing clauses in many contexts.

Finally in terms of the practical efficiency of a patent holder-led model for ethical licensing clauses, if the purpose of a licensing clause is to discourage or limit specific uses of a technologies which raise ethical issues, and particularly where there is agreement within the scientific community that such uses

[46] Aisling McMahon, 'Biotechnology, Health and Patents as Private Governance Tools: The Good, the Bad and the Potential for Ugly?' (2020) 3 Intellectual Property Quarterly 161.

[47] www.broadinstitute.org/history-leadership/founders accessed July 2021.

[48] www.broadinstitute.org/strategic-alliances-and-partnering accessed July 2021.

[49] Jorge L Contreras and Jacob S Sherkow, 'CRISPR, Surrogate Licensing, and Scientific Discovery' (2017) 355 (6326) Science 698.

[50] For a discussion of the role of financial drivers in the corporate decision-making context around patent licensing, see: Aisling McMahon and Edana Richardson, 'Patents, Healthcare & Engaged Shareholders: A Pathway to Encourage Socially Responsible Patent Use?' (2021) Legal Studies (FirstView) 1. doi:10.1017/lst.2021.49

should be discouraged,[51] licensing clauses would likely need to be adopted by many rightsholders for this to have a significant blocking effect on the use of a technology in a national or international sense. This is because often many rightsholders will have patents which are relevant to a technological field or context.[52] For instance, there are many relevant rightsholders in the CRISPR-Cas9 context.[53] Thus, a licensing clause adopted by one rightsholder, where other rightsholders do not adopt similar clauses, could mean that clause is less effective, as third parties could seek a work-around by, for example, moving operations to another jurisdiction where no patents apply, or where the relevant rightsholder(s) may differ.[54]

In short, *patent holder-driven* ethical licensing clauses in some contexts can give rise to questions in terms of accountability of rightsholders in the adoption and design of such clauses and around the efficiency of the use of licensing clauses by individual patent holders.

3.3 Ethical Licensing Conditions as a Complementary Regulatory Tool

Nonetheless, ethical licensing clauses could instead be adopted as part of a broader regulatory or governance approach to how emerging technologies are developed.[55] In this context, for instance, ethical licensing clauses could

---

[51] An example of this are concerns around the use of gene editing technologies for clinical reproductive purposes.

[52] A recent example showing this is a patent landscape study for mRNA patents over COVID-19 vaccines – see: Mario Gaviria and Burcu Kilic, 'A network analysis of COVID-19 mRNA vaccine patents' (2021) 39 Nature Biotechnology 546.

[53] Jacob S Sherkow, 'The CRISPR Patent Landscape: Past, Present, and Future' (2018) The CRISPR Journal 5; Paramita Ghosh, 'Patent Landscape of CRISPR/Cas' in Anjanabha Bhattacharya, Vilas Parkhi and Bharat Char (eds) *CRISPR/Cas Genome Editing. Concepts and Strategies in Plant Sciences* (Springer 2020).

[54] In the context of a discussion of different rightholders/patents operating depending on the jurisdiction, see: Jacqueline Martin-Laffon, Marcel Kuntz, and Agnes E Ricroch, 'Worldwide CRISPR Patent Landscape Shows Strong Geographical Biases' (2019) 37 Nature Biotechnology 613.

[55] See also discussion of this in the context of gene editing in: WHO Expert Advisory Committee on Developing Global Standards for Governance and Oversight of Human Genome Editing, Human Genome Editing a Framework for Governance (July 2021); Duncan Matthews, Abbe Brown, Emanuela Gambini, Aisling McMahon, Timo Minssen, Ana Nordberg, Jacob S. Sherkow, Jakob Wested, and Esther van Zimmeren, 'The Role of Patents and Licensing in the Governance of Human Genome Editing: A White Paper' (30 July 2021) Queen Mary Law Research Paper No. 364/2021; Shobita Parthasarathy, 'Use the Patent System to Regulate Gene Editing' (2018) 562(7728) Nature 486. More generally, see: Aisling McMahon, 'Biotechnology, Patents and Licensing for "Ethical Use": A Regulatory Opportunity? (Working Paper

be recommended by national bioethics councils or national governments for adoption by rightsholders over specific technological uses. This would also tie in with the jurisdictional nature of patents – as noted above, patents and rightsholders may differ depending on the jurisdiction, as patents are generally national rights.[56] Thus, one could envisage a system whereby a national bioethics body, for example, in consultation with groups and stakeholders such as the national patent office, intellectual property experts, scientific personnel and so on, could develop recommendations on licensing conditions where concerns arose. This could provide a useful avenue to recommend restrictions on uses of patented technologies where concerns arose.

Alternatively, for greater international reach and greater effectiveness in blocking contentious uses of emerging technologies more broadly, such clauses could be recommended at an international level. For instance, in the health context these could be considered and recommended by entities such as the World Health Organization (WHO),[57] or via a partnership of entities working together, such as the WHO working alongside relevant international intellectual property entities such as the WIPO or WTO,[58] for adoption by rightsholders or within and by patent pools.[59] The precise operational modal-

2021); Aisling McMahon (2019) "Patent licensing for 'ethical' use: A missed regulatory opportunity?", ATRIP Conference Presentation, Vanderbilt University, Nashville, Tennessee , 26 August 2019.

56 This is discussed in: Duncan Matthews, Abbe Brown, Emanuela Gambini, Aisling McMahon, Timo Minssen, Ana Nordberg, Jacob S. Sherkow, Jakob Wested, and Esther van Zimmeren, 'The Role of Patents and Licensing in the Governance of Human Genome Editing: A White Paper' (30 July 2021) Queen Mary Law Research Paper No. 364/2021.

57 Aisling McMahon, 'Biotechnology, Patents and Licensing for "Ethical Use": A Regulatory Opportunity?' (Working Paper 2021).

58 In the human genome context, see: Duncan Matthews, Abbe Brown, Emanuela Gambini, Aisling McMahon, Timo Minssen, Ana Nordberg, Jacob S. Sherkow, Jakob Wested, and Esther van Zimmeren, 'The Role of Patents and Licensing in the Governance of Human Genome Editing: A White Paper' (30 July 2021) Queen Mary Law Research Paper No. 364/2021, 50 available at: https://ssrn.com/abstract=3896308 accessed 31 July 2021. For a recent proposal which proposes an amendment of TRIPS to facilitate the role of the WTO system via a WTO ethics committee being able to make recommendations, which could include ethical licensing clauses, see: Oliver Feeney, Julian Cockbain and Sigrid Sterckx, 'Ethics, Patents and Genome Editing: A Critical Assessment of Three Options of Technology Governance' (2021) Frontiers in Political Science (21 September 2021).

59 See in the human genome editing context: Duncan Matthews, Abbe Brown, Emanuela Gambini, Aisling McMahon, Timo Minssen, Ana Nordberg, Jacob S. Sherkow, Jakob Wested, and Esther van Zimmeren, 'The Role of Patents and Licensing in the Governance of Human Genome Editing: A White Paper' (30 July 2021) Queen Mary Law Research Paper No. 364/2021, 50.

ities of how such ethical licensing approaches would be developed and fit within the broader regulatory context are outside the scope of this chapter, and are considered by this author elsewhere.[60] Instead, the focus here is on making the broader general argument providing a justification for why ethical licensing clauses to discourage contentious uses of patented technologies could be a useful complementary regulatory tool and warrant greater scrutiny. In this context, aside from the rationale for adopting such clauses within patent licenses in terms of ensuring alignment with patent's social bargain, five key benefits of using ethical conditions at patent licensing stage as part of a broader package to assist in the regulation of emerging patented technologies can be identified.

First, as technologies have developed – particularly in the context of emerging biotechnologies – they increasingly push the boundaries of what should or should not be encouraged in terms of scientific development, and can give rise to ethical concerns. In such cases, different uses of an emerging technology under patent may arise. Many of these uses may be considered beneficial to society. However, some may give rise to contention within the scientific community. A key recent example of this is in the context of human genome editing technologies.[61] In this context, many uses of genome editing technologies are highly beneficial to society, including for research and development of healthcare and knowledge of existing diseases. However, certain aspects of genome editing technologies, particularly in relation to germline clinical uses of human embryos for reproductive purposes, pose significant ethical concerns. Accordingly, there have been calls for a global moratorium on clinical uses of human genome editing.[62] In such contexts, where a technology

---

[60] Aisling McMahon, 'Biotechnology, Patents and Licensing for "Ethical Use": A Regulatory Opportunity?' (Working Paper 2021); Aisling McMahon, "Patent licensing for 'ethical' use: A missed regulatory opportunity?" ATRIP Conference, Vanderbilt University, Nashville, Tennessee (26 August 2019).

[61] The regulation of genome editing has already been subject to several national and international reports, including: National Academy of Medicine, *National Academy of Sciences, and the Royal Society, Heritable Human Genome Editing* (The National Academies Press 2020) https://doi.org/10.17226/25665 (US context); Nuffield Council on Bioethics, *Genome Editing and Human Reproduction: Social and Ethical Issues* (Nuffield Council on Bioethics 2018). The World Health Organization has also recently published a report which examines the governance of human genome editing and highlights the role that patents and licences could play in this context, see: WHO Expert Advisory Committee on Developing Global Standards for Governance and Oversight of Human Genome Editing, *Human Genome Editing: A Framework for Governance* (July 2021) available at: www.who.int/publications/i/item/9789240030060 (accessed 31 July 2021).

[62] Eric Lander, Françoise Baylis, Feng Zhang et al, 'Adopt a Moratorium on Heritable Genome Editing', Nature 13 March 2009) available at: www.nature.com/articles/d41586-019-00726-5 accessed 31 July 2021.

has multiple potential uses but concerns arise only over some specific uses, the patent licensing system could be used as a vehicle to include licensing clauses restricting contentious use(s) of that technology where there is consensus within the national or international community (depending on what level that clause were recommended for adoption) around the contentious nature of such uses.

Second, and relatedly, in such contexts this use of ethical licensing clauses rather than denial of a patent grant over the technology more generally (as this is often the main avenue where ethical issues within the patent system have been discussed to date)[63] is particularly appropriate. This is because in such cases many of the uses of that technology may be beneficial for society, and hence something which should be encouraged via a patent grant. However, some specific uses of that technology may be sought to be discouraged or prohibited. For example, as noted, genome editing technologies have many beneficial applications, but concerns also arise around specific uses of such technologies, for example, for gene drives. In such cases a patent licensing condition restricting that use could allow for the technology to gain some of the benefits of patent protection, while also discouraging specific uses of that technology that are ethically contentious.[64] Thus, ethical licensing clauses

---

[63] There is an extensive body of literature on the role of morality/ordre public provisions at patent grant stage in Europe – see generally: Derek Beyleveld and Roger Brownsword, *Mice, Morality and Patents* (Common Law Institute of Intellectual Property, 1993); Peter Drahos, 'Biotechnology Patents, Markets and Morality' (1999) 21(9) European Intellectual Property Review 441; Oliver Mills, *Biotechnological Inventions: Moral Restraints and Patent Law* (Revised Edition) (Ashgate Publishing Limited 2010); Aurora Plomer and Paul Torremans (eds), *Embryonic Stem Cell Patents: European Law and Ethics* (Oxford University Press 2009); Margo Bagley, 'Patent First, Ask Questions Later: Morality and Biotechnology in Patent Law' (2003–4) 45 William and Mary Law Review 469; Lionel Bently and Brad Sherman, 'The Ethics of Patenting: Towards a Transgenic Patent System' (1995) 3 Medical Law Review 275; Rainer Moufang, 'Patenting of Human Genes, Cells and Parts of the Body? The Ethical Dimensions of Patent Law' [1994] 25(4) IIC 487. For more recent discussions, see: Aisling McMahon, *The Morality Provisions in the European Patent System: An Institutional Examination* (PhD Thesis, University of Edinburgh 2016); Ana Nordberg, 'Patents, Morality and Biomedical Innovation in Europe: Historical Overview, Current Debates on Stem Cells, Gene Editing and AI, and de lege ferenda Reflections' in Daniel Gervais (ed) *Fairness, Morality and Ordre Public* (Edward Elgar Publishing 2020); Justine Pila, 'Adapting the Ordre Public and Morality Exclusion of European Patent Law to Accommodate Emerging Technologies' (2020) 38 Nature Biotechnology 555; Karen Walsh and Naomi Hawkins, 'Expanding the Role of Morality and Public Policy in European Patent Law' in Paul Torremans (ed.) *Intellectual Property and Human Rights* (4th ed Wolters Kluwer 2021).

[64] This scenario described contrasts with a scenario whereby there is an objection to the commercial exploitation of the technology by patent more generally, in which

provide the ability to tailor conditions, as opposed to introducing a blanket denial of a patent.

Moreover, in such contexts, if the aim of an initiative within the patent system is to restrict a use of a technology under patent, refusal of a patent grant would not guarantee this and could in fact have the opposite effect. This is because the denial of a patent – based on the morality provisions, for example[65] – has the effect that patents are not available on that technology in that jurisdiction. Given the negative nature of the patent right, refusing a patent simply means the patent holder has no right to stop others from using the technology. However, denial of a patent does not limit others' use of the technology; in fact, refusal of a patent could also potentially increase uses of a technology, as there can be no threat of patent infringement if no patent applies. Hence, if the regulatory aim is to limit specific uses of a technology, denial of a patent at grant stage cannot directly achieve this: this is the so-called futility or inefficiency argument, which is often levelled at the patent system in discussions on patents' role within the broader regulatory context.

Third, ethical conditions within patent licensing used in this manner act as a useful complement to broader regulatory approaches for emerging biotechnologies and other emerging technologies.[66] As noted, the patent office is often the first point at which an invention is examined by an external scientific community beyond the inventors.[67] Thus, the patent system can provide an important forum within at which an ethical issue with an emerging technology may be identified, as at this point the technology is scrutinised by external scientific personnel (prior to this, the technology must not be in the public domain, generally to obtain patent protection). Where identified, potential ethically contentious uses of patent claimed technology could be referred to international bodies or ethics councils to consider and assess whether an ethical licensing restriction might be useful or could be recommended.

Fourth, it is likely that in some contexts the possible uses of a technology will develop over time: as that technology is used by others in the field, more may become known about it. Using the patent licensing stage to incorporate conditions as to how a technology is used could provide a useful stopgap to

---

case denial of a patent grant would arguably be more appropriate and warranted under Article 53(a) EPC, or Article 6(1) of the Biotechnology Directive 98/44EC.

65 In the European context, these provisions are contained in: Article 53(a) EPC or Article 6 Biotechnology Directive 98/44EC.

66 See, also Shobita Parthasarathy, 'Use the Patent System to Regulate Gene Editing' (2018) 562(7728) Nature 486.

67 For an invention to be patentable it must meet novelty criteria under the relevant patent laws, so often the first time that invention is examined by the scientific community (outside of the inventors) will be at the patent examination stage.

allow conditions to be recommended and then be adopted by rightsholders as a temporary measure when ethical issues are identified either at grant stage or at a later point in time. This would also allow time for other relevant regulatory bodies to address such issues, while sending a message to the broader scientific community on certain uses which may be deemed ethically problematic or contentious. Use contrary to that ethical licensing clause would render the third party either in breach of the licence (if they were a licensee) or potentially in infringement of the patent (if no licence applied), and penalties would apply for both scenarios, which could act as a deterrent against the restricted use of the technology. Adopting such conditions at patent licensing stage could be done in a flexible manner,[68] as recommendations in relation to licensing conditions could be modified more becomes known about the technology.

Finally, such clauses are contractual provisions and could be adopted in an expedient fashion within the patent licence by the licensor and the licensee.[69] The adoption of patent licensing clauses is likely to be quicker than the adoption of other regulatory measures, such as prohibitions on use of a technology more generally, which may require adoption via national legislation or an international treaty (depending on the geographical scope of such provisions), which can take many years to adopt. In contrast, if, for example, patent licensing conditions were recommended for adoption by an international organisation such as the WHO, rightsholders could include these in the patent licence in an expedient fashion, as it would require only a contractual provision. Moreover, if these conditions were recommended on a voluntary basis by an entity such as the WHO, it is likely that rightsholders may be inclined to adopt them due to the potential risk of reputational fall-out if they ignored such a recommendation especially, if the purported ethical concern related to the technology later arose.[70] Where concerns around uses of an emerging technology arose, this would provide a means to flag and pre-empt them either as a complement to other regulatory approaches, or as a temporary measure prior to other regulatory systems adopting comprehensive regulation. Scientists themselves, including developers of technologies, are increasingly expressing concerns

---

[68] The flexibility advantage associated with ethical licensing clauses, albeit in the context of patent holder-led approaches, is discussed in: Oliver Feeney, Julian Cockbain and Sigrid Sterckx, 'Ethics, Patents and Genome Editing: A Critical Assessment of Three Options of Technology Governance' (2021) Frontiers in Political Science (21 September 2021).

[69] Ibid – in the context of patent-holder led ethical licensing approaches, Feeney et al discuss the advantage of 'speed' in terms of their adoption associated with such clauses.

[70] This is developed further in: Aisling McMahon, *Biotechnology, Patents and Licensing for 'Ethical Use': A Regulatory Opportunity?* (Working Paper 2021).

about the potential unforeseen uses a technology could have and potential harms to society that could arise which they did not intend or foresee[71] thus, this avenue may also likely be supported by inventors themselves.

4. A CASE FOR GREATER USE AND SCRUTINY OF ETHICAL CONSIDERATIONS IN THE LICENSING OF HEALTH-RELATED TECHNOLOGIES: EQUITABLE ACCESS TO PATENTED HEALTH TECHNOLOGIES

This section outlines six key reasons why ethical considerations related to how patent rights are used over health technologies should be given greater consideration on a more routine basis at the patent licensing stage, and specifically, why there needs to be greater consideration of how use of the *patent right itself* can affect downstream accessibility of patented technologies and how licensing clauses could be used to ameliorate this.[72]

First, as discussed above, patents – unlike other forms of intellectual property rights – are not automatic rights granted to the inventor of a technology. Instead, patents are limited exclusivity rights that are granted in return for that inventor delivering technologies which meet certain key patentability criteria, namely, novelty, inventive step and technical application.[73] Accordingly, patent rights are not an automatic entitlement, their grant is always conditional. Relatedly, if one again considers the patent social bargain rationale if in theory the patent right is offered as a limited exclusivity right in return for delivering an invention which is of benefit to society. That being the case, there is arguably strong justification for limiting the scope of rightsholders' discretion over

[71] For example, Jennifer Doudna, one of the leading scientists involved in the development of CRISPR technologies, has described having nightmares about gene editing technologies being used for nefarious purposes, and the potential harms to society that could arise – as discussed in: Jennifer Doudna and Samuel Sternberg, A Crack in Creation: Gene Editing and the Unthinkable Power to Control Evolution (Houghton Mifflin Harcourt 2017).

[72] In the human gene patent context, see: Sarah E. Ali Khan and Richard Gold, 'Gene Patents Still Alive and Kicking: Their Impact on Provision of Genetic Testing for Long QT Syndrome in the Canadian Public Health-Care System' (2017) 19 Genetics in Medicine 1253; Sarah Ali-Khan and E. Richard Gold, 'Contracting to Counter Gene Patents – A 21st Century Solution to Access and Innovation' (Petrie Flom Blog, 22 May 2017) http://blog.petrieflom.law.harvard.edu/2017/05/22/contracting-to-counter-gene-patents-a-21st-centurysolution-to-access-and-innovation/ (accessed September 2021).

[73] See Arts 54, 58 and 57 of the EPC, as amended. See also Art 27(1) TRIPS Agreement, as amended. This is the terminology used in the European context: note that the terminology for such terms can differ across jurisdictions.

patent use, which could include recommending or mandating conditions on the licensing of certain technologies where the public interest, including human health, requires this.

Second, and relatedly, attaching conditions to patent rights as a form of intangible property right to facilitate public access or use of an invention (in some contexts) would not be unprecedented if we consider that in the tangible property context it is common that conditions on tangible property rights apply. For example, water rights can subsist over real property to allow third parties who occupy surrounding lands access to water as an essential resource;[74] similarly, rights of way can exist over land. The property owner in such contexts has a valid property right, but that does not mean such rights cannot be fettered in this way by providing entitlements to third parties. Similarly, one could make the case that particularly in the health context – and especially for essential health technologies, which could be considered a key health resource – where such technologies are patented, licensing conditions facilitating reasonable access for specific groups who may otherwise be unable to access such patented technology (such as LMICs, due to affordability or other issues) would be entirely reasonable and appropriate, and in some cases may need to be mandated.

Third, this chapter argues that the vast changes in technologies since the development of early patent systems, and even since the advent of regional patent systems such as the European Patent Convention in the 1970s, offer an additional reason why we need to question how rightsholders' control is exercised post-grant, and specifically, why greater scrutiny is needed over how ethical issues pertaining to access to technologies could be addressed via licensing conditions. Traditionally, patents were developed in the context of industrial or mechanical inventions, such as mechanical engine parts. In comparison to many industrial-era technologies, lack of access to patentable contemporary health technologies, such as a medicine, or medical device, can have significant impacts on human health, with the potential for fatal consequences.[75] Furthermore, where patents hinder access to such technologies this can directly conflict with human rights at stake, such as the right to life and

[74] For a discussion of water rights, see: Jill Robbie, *Private Water Rights* (Edinburgh: Edinburgh Legal Education Trust, 2015); see also: Aisling McMahon, 'Biotechnology, Health and Patents as Private Governance Tools: The Good, the Bad and the Potential for Ugly?' (2020) 3 *Intellectual Property Quarterly* 161, 164.

[75] It is conceded that lack of access to many other technologies aside from health technologies can have significant repercussions for human life and wellbeing, and this argument should not be taken to mean that scrutiny over licensing of other technologies is not needed; rather, such issues are beyond the scope of this current chapter.

right to health, which are protected by various international treaties.[76] Owing to the potential impacts for health arising from how patent rights are used over such technologies, this adds further impetus for why a reconsideration of the role of ethical issues at patent licensing stage is needed.

Fourth, and relatedly, some health technologies have a peculiar relationship with the human body, wherein the line between the body part and the technology developed and can become blurred. For example, isolated human genes are still patentable in Europe.[77] In such contexts, the traditional line between the human body, which is unpatentable, and technology, which is patentable, becomes blurred in such contexts.[78] Arguably, this blurring of boundaries lends further support to why, in some cases, conditions should be imposed over how the exclusive rights provided by a patent over such 'technologies' (where they are patented) are used by rightsholders to ensure reasonable downstream access to such technologies. Moreover, granting a patent over such technologies with no conditions on how that patent is used can be highly problematic, given the potential impacts which could arise for bioethical interests and human rights at stake. For example, patents over such technologies and how such patents are used can have significant implications for interests such as human dignity (if, for example, an element of the body is deemed patentable), bodily integrity and for how individuals access information or treatment.[79] Again, this suggests a role either for limiting what patents are granted over such technologies or for

---

[76] The right to health is protected by several international treaties, including Article 25 Universal Declaration of Human Rights (UDHR) and Article 12 of the International Covenant on Economic, Social and Cultural Rights. For a discussion of the right to health and the patent context, see: Duncan Matthews, 'The Right to Health and Patents' in C Geiger (ed.) *Research Handbook on Human Rights and Intellectual Property* (Edward Elgar Publishing 2015) 496.

[77] In Europe the human body itself is not patentable, however, isolated elements of the body may be patentable: Article 5, Biotechnology Directive.

[78] Other examples of this include technologies which rely on information or elements from the body to operate, such as genetic testing, or chimeric antigen receptor T-cell therapy (CAR-T). For a discussion of patentability in the CAR-T context, see: Luis Gil Abinader and Jorge L Contreras, 'The Patentability of Genetic Therapies: CAR-T and Medical Treatment Exclusions Around The World' (2019) 34(4) American University International Law Review 705.

[79] This is discussed in: Aisling McMahon, 'Covid-19, Patents & Healthcare: The Need for A (Bio)ethics Space within Patent Law' *Journal of Medical Ethics Blog* (16 April 2020) available at https://blogs.bmj.com/medical-ethics/2020/04/16/covid-19-patents-healthcare-the-need-for-a-bioethics-space-within-patent-law/ accessed 3[illegible] July 2021; See also: TT Arvind and Aisling McMahon, 'Commodification, control, and the contractualisation of the human body' in E Bertrand, M-X Catto and A Mornington (eds) *The Limits of the Market: Commodification of Nature and Body* (Mare & Martin 2020). The role of bioethics within patent law, and the bioethical implications posed by patent grant and use for technologies related to the human body will be investigated

placing conditions on the patent related to how it is licensed, to address ethical issues including access concerns that may arise over how a patent right over such technologies is used (or both).

Fifth, prior to the adoption of the TRIPS Agreement, States had the ability to vary patent laws over health technologies, including by abolishing patents for certain technologies such as pharmaceuticals. As a result, many States had no patents for pharmaceuticals or abolished such systems, particularly in the 1960s and 1970s, and this allowed greater legal scope for such States to produce generic medicines.[80] Since the adoption of the TRIPS Agreement this is no longer possible, as Art 27 of the TRIPS Agreement states that patents must be available for all fields of technology. As this provision is applicable in all World Trade Organisation States, this limits discretion for WTO States over what types of patents are granted in the health context. Accordingly, given that the fields of technologies which can be patented cannot be varied in WTO States, how patents are used over certain technologies, such as health technologies, warrants greater scrutiny given the impact that uses of patent rights over health technologies can have for human health and life, and particularly for LMICs, who are often most affected by access issues related to patented health technologies. In such contexts it is entirely plausible that tailored patent licensing clauses could be adopted on a more routine basis which would ensure broader access to patented technologies for groups which may otherwise be unable to access them, such as by introducing a clause that access would be provided on a non-profit basis for LMICs.

Sixth, it is important to highlight that access issues related to patent use cannot always be best addressed by the *refusal of a patent grant* per se (that is not to say that in some cases refusal of patent grant may not be wholly warranted and needed). However, in some cases it may be more appropriate to take ethical considerations (in terms of access issues) into account at licensing stage rather than refuse a patent. Similar to the use of ethical licensing clauses, this would allow some of the benefits of patent grant to go to the rightsholder,[81] but

---

further as part of the 'PatentsInHumans', European Research Council funded Starting Grant project (PI: Prof McMahon) which will commence in Autumn 2022.

[80] See: Siva Thambisetty, Aisling McMahon, Luke McDonagh, Hyo Yoon Kang and Graham Dutfield, The TRIPS Intellectual Property Waiver Proposal: Creating the Right Incentives in Patent Law and Politics to end the COVID-19 Pandemic (24 May 2021) LSE Legal Studies Working Paper No. 06/2021.

[81] In this context the availability of a patent is sometimes argued to be a useful incentive for the development of a technology, and it is sometimes argued that the absence of any patent right over a technology could potentially in some (but not all) cases hinder the development of health technologies. However, the precise role of patents in incentivising technologies can differ, and this is contested in the literature. Such issues are beyond the scope of this current chapter.

in a manner which curbs the potential ethical issues arising, by adopting ethical access licensing clauses.

The foregoing six reasons support the case for why ethical issues in terms of access (and how these may be ameliorated by licensing conditions) should be considered more widely in the licensing of patented health-related technologies to ensure that downstream access to such technologies is not unreasonably hindered. Discussion of the specific system by which such licensing conditions could be adopted within patent agreements is beyond the scope of this current chapter. The purpose here is to build this justification, and to call for greater scrutiny of ethical issues around how patented health technologies are licensed in terms of facilitating broader access to such technologies.

Nonetheless, in this context, similarly to the ethical licensing context, it must be noted that patent holder-led approaches to licensing clauses facilitating ethical access licensing clauses also give rise to issues. In particular, there are often likely to be limited (financial) incentives for rightsholders to adopt such clauses. Patent holders operating on a for-profit basis could instead incur financial detriment by introducing such clauses, as in many cases they may need to lower the price of patented technologies to ensure accessibility of such technologies for particular groups. An additional issue is the power asymmetry that often exists between rightsholders (who may be large commercial entities) and those seeking access to patented health technologies, including patients, hospitals, and so on.[82] Such rightsholders often will have more financial resources to cover legal costs and may have more time to enable them to sit out legal proceedings or to make ongoing challenges/appeals, which may discourage third parties such as patients or hospitals challenging patents or terms within a licensing agreement. This asymmetry must be borne in mind in any conversation around the role of licensing clauses to facilitate broader access to patented technologies.[83]

For ethical access licensing clauses to be more embedded within the health innovation landscape, one avenue which should be considered is the role of intermediaries who can leverage rightsholders to adopt such clauses. 'Intermediaries' is used in this context to refer to any actor that has a resource which entities developing technologies may need access to, that could in turn

---

[82] Aisling McMahon, 'Gene Patents and the Marginalisation of Ethical Issues' (2019) 41(10) European Intellectual Property Review 608–20.

[83] For a discussion of this and a proposal on how to counter it in the gene patent context by greater governmental intervention see: Sarah Ali-Khan and E. Richard Gold, 'Contracting to Counter Gene Patents – a 21st Century Solution to Access and Innovation' (Petrie Flom Blog, 22 May 2017) available at: http://blog.petrieflom.law.harvard.edu/2017/05/22/contracting-to-counter-gene-patents-a-21st-century-solution-to-access-and-innovation/ accessed September 2021.

be leveraged to mandate the adoption of ethical access licensing clauses. In this context, examples of relevant actors include research funders; public institutions where research takes place, such as academic institutions; and biobanks whose biological specimens are required in the development of such technologies. Such entities could play a significant role in normalising the use and adoption of ethical access licensing clauses, by leveraging their power in terms of placing ethical licensing conditions on access to the resources they hold (for example, via research funding agreements or in agreements on material transfer of biological samples) to recommend or preferably mandate that those using such resources to agree to licensing conditions which account for potential ethical issues arising relating to patent use.

Examples of such clauses or guidance in relation to these types of licensing principles, are already evident. For instance, in the university context, in 2007 12 universities collaborated to draft the 'AUTM Nine Points to Consider in Licensing University Technology',[84] which has more than 100 university signatories.[85] These guidelines include as point 5 the need to 'ensure broad access to research tools' and as point 9 that entities will '[c]onsider including provisions that address unmet needs, such as those of neglected patient populations or geographic areas, giving particular attention to improved therapeutics, diagnostics and agricultural technologies for the developing world'. In the context of point 9, the guidelines specifically refer to the 'social compact with society' which universities have and state that accordingly, universities 'have a specific and central role in helping to advance knowledge in many fields and to manage the deployment of resulting innovations for the public benefit. In no field is the importance of doing so clearer than it is in medicine.' As part of this role, the document suggests:

> We have a responsibility to try to alleviate [the failure to prevent or treat disease], including finding a way to share the fruits of what we learn globally, at sustainable and affordable prices, for the benefit of the world's poor. There is an increased awareness that responsible licensing includes consideration of the needs of people in developing countries and members of other underserved populations.[86]

---

[84] AUTM describes itself as a 'non-profit leader in efforts to educate, promote and inspire professionals to support the development of academic research that changes the world and drives innovation forward. Our community is comprised of more than 3,000 members who work in more than 800 universities, research centers, hospitals, businesses and government organizations around the globe.' See: https://autm.net/about-autm/who-we-are/ accessed 31 July 2021.

[85] https://autm.net/about-tech-transfer/principles-and-guidelines/nine-points-to-consider-when-licensing-university accessed 31 July 2021.

[86] www.autm.net/AUTMMain/media/Advocacy/Documents/Points_to_Consider.pdf at 8, accessed 31 July 2021.

Furthermore, a recent example of socially responsible licensing clauses is the 'Netherlands Federation of University Research Medical Centers principles for socially responsible licensing' (NFU guidelines) adopted in June 2019. These guidelines were adopted following work by a national working group in the Netherlands involving a range of collaborators and stakeholders, and a broad consultation of the field, which was commenced due to the Minister for Medical Care and Sport's request to consider the potential for socially responsible licensing.[87] Principle 10 in this document states: 'Licences ensure that the price-setting of the final products and/or services does not endanger accessibility'.[88] It further suggests that:

> When arranging a licence, it can therefore be agreed that the partner will endeavour to make a reasonable commercial effort to ensure that the final price of the product or service will not hinder its availability in a particular market. The criterion to determine what is acceptable depends on the context at the time that the product or service is marketed.[89]

Such principles again highlight a pathway to encourage licensing of technologies in a manner that aligns with facilitating broader access for technology users.

However, in many cases licensing guidelines relating to access issues may be non-binding in nature, or drafted in an open manner so that the scope of third parties' obligations is not always clear-cut.[90] Such issues may in turn impact how universities, biobanks or other funders, and the third parties they work with, adopt and enact guidelines (where applicable) in practice. Moreover, even if a university or funder were to adopt clauses seeking to encourage ethical downstream access to patented technologies, whether such clauses would be enforced in practice would be questionable, as there may be limited incentives or resources for such intermediary entities to monitor how licensing clauses/guidance is enacted or complied with by third parties. It is also often not clear what penalties may apply for failure of third parties to facilitate such access, and how these would be enforced in practice. Such issues require further scrutiny and investigation.

---

87 www.nfu.nl/sites/default/files/2020-08/19.4511_Ten_principles_for_Socially_Responsible_Licensing_v19-12-2019.pdf at 4, accessed 31 July 2021.

88 Ibid.

89 Ibid at 13.

90 For a discussion of issues which can arise with the open-textured nature of clauses around IP in the biobank context, see: Aisling McMahon (2016) 'Patents, Human Biobanks and Access to Health Benefits: Bridging the Public–Private Divide' in Jessica Lai and Antoinette Maget Dominicé (eds) *Intellectual Property and Access to Im/Material Goods* (Edward Elgar Publishing 2016) section 5.3.

In short, a deeper conversation around the role of ethical licensing clauses to facilitate broader access to patented health technologies, which includes scrutiny of the power asymmetries arising and the avenues to alleviate this, is warranted. Intermediary bodies such as funders, biobanks[91] and universities play an important role as the resources they hold provide one means to leverage rightsholders into adopting ethical approaches to licensing of patented technologies. Moreover, if such clauses were adopted by all major research funders, and if clauses like this were routinely enforced, this could encourage the development of an industry norm which encourages broader reflection on and adoption of ethical access licensing clauses for patented health technologies. Thus, it is vital that such stakeholders are part of a broader conversation around how to embed ethical considerations around access to health technologies at patent licensing stage. Similarly, normalising mandatory interventions with rightsholders' discretion, such as via compulsory licensing where access issues arise, could also prove useful to encourage a broader cultural change in licensing practice which seeks to facilitate broader equitable access to patented technologies in the health context.[92]

5. CONCLUDING THOUGHTS

This chapter has argued that, for patented biotechnologies and health technologies related to the human body, there needs to be far greater scrutiny of how ethical issues are considered at the patent licensing stage, both in terms of the potential for patent licensing clauses to be harnessed and used in a manner which complements existing regulatory approaches to emerging technologies by restricting/discouraging ethically contentious uses of patented technologies via ethical licensing clauses and by countering the potential negative repercussions of how patent rights can be used to hinder access to downstream technologies, via ethical access licensing clauses. In effect, alongside the economic function of a patent, patent rights bestow upon rightsholders a significant governance function over patented technologies which allows them to act as gatekeepers over how the technology subject to patent is used and accessed for

[91] See also: Hermann Garden, Naomi Hawkins and David Winickoff, 'Building and Sustaining Collaborative Platforms in Genomics and Biobanks for Health Innovation' (2021) OECD Science, Technology and Industry Policy Papers, No. 102 (2021) Éditions OCDE, Paris, https://doi.org/10.1787/11d960b7-en; Naomi Hawkins, 'Biobanks and Collaborative Platforms in Genomics – What Role for Patents?' (Patent Scholars Network, 28 April 2021).

[92] See also: Aisling McMahon and Edana Richardson, 'Patents, Healthcare & Engaged Shareholders: A Pathway to Encourage Socially Responsible Patent Use?' (2021) Legal Studies (FirstView) 1. doi:10.1017/lst.2021.49

the duration of the patent term. Given the nature of contemporary patentable health technologies, the impact of rightsholders' decisions regarding how they use such patent rights can have significant implications for human health and life. Thus, the discretion given to rightsholders in such contexts, including their role in dictating terms of patent licences over such technologies, warrants much deeper investigation.

In this context, to date much of the discussion of ethical issues related to patent rights focused primarily on the patent grant stage – and in the European context, particularly on the role of morality provisions in refusing patents based on morality/*ordre public* concerns around the commercial exploitation of a patented invention. While such discussions are important, there also needs to be greater consideration of the role of patent licensing as an avenue to engage with ethical issues arising from both the use of the patent right and the use of the technology under patent. This chapter has focused on building the general case for why ethical considerations should be taken into account at patent licensing stage under both categories, including the potential benefits of doing so. It has also highlighted some of the key challenges and issues which would need to be addressed – such as the power asymmetries at play, the (dis) incentives to engage with ethical issues at licensing stage and issues around enforcement of such clauses – if ethical considerations are to be accounted for at patent licensing stage in a meaningful manner which aligns with the broader public interests at stake. More generally, this chapter has called for a much deeper interdisciplinary conversation across legal, ethics, scientific and health disciplines on the role of ethics within patent licensing. This conversation is urgently needed if humanity is to realise in full the benefits of emerging technologies while also steering uses of such technologies in an ethically prudent manner.

9. Patents and the public interest: a multi-faceted approach

Naomi Hawkins

1. INTRODUCTION

This book has explored important questions for the discipline of patent law arising from developments in biotechnological innovation. Patent law has always been focused on new and emerging areas of technology; the requirement of novelty is such that the subject matter of a patent has always been 'new'. But the quantitative novelty required for a successful patent application does not always set a high bar, and as a result, many of the patents granted by a patent office in any year will not fit the description of being an 'emerging' technology. Instead, an emerging technology brings a step-change development – such as the biotechnologies discussed in this book. Fields such as non-invasive prenatal testing, bioprinting, the interaction of the internet of things and the body, and artificial intelligence are all driving biotechnology in ways that would have been unimaginable decades ago.

Such developments necessitate consideration of the question of whether existing patent law principles are fit for purpose. Are existing principles of law sufficiently flexible for their application to new areas of technology to be workable? And, equally importantly, does the existing scheme of laws result in a fair and appropriate outcome? As discussed in the Introduction, and as explored throughout this book, patents are only justified if they serve the public interest. They are intended to function to incentivise innovation and to disclose the invention to the public – to serve the public interest as quid pro quo for the private gain associated with the monopoly right granted.

The patent system depends on a balance between private rights and the public interest, with that balance achieved through the award of the patent monopoly right only where the information revealed in the patent application provides that corresponding benefit to the public. The bargain is central to the operation of patent law, but whether this balance is struck at the right level – whether the public interest is adequately safeguarded – is debateable. This question has been at the core of the consideration of this book.

The law provides many ways through which private rights and the public interest are balanced. Limitations on patentable subject matter, such that patents are only for inventions (in Europe) or manners of new manufacture (in Australia) or where products of nature are excluded matter (USA), serve to exclude certain types of subject matter from the scope of the patent system, in ways that may safeguard the public interest. The lines drawn in relation to the application of the criteria of novelty, inventive step, industrial application and insufficiency also have important implications for the breadth of what is or is not patentable, with an implicit link to the public interest.[1] Other aspects of the patent system, such as the *ordre public* and morality exceptions,[2] have a more explicit relationship to the public interest. Patent exploitation also has important implications for the public interest. The free market approach and freedom of contract result in a very narrow scope for the public interest, but the importance of compulsory licensing, competition law controls and the use of contractual bargaining power to safeguard the public interest are all fruitful areas of consideration.

Finally, and significantly, questions of non-compliance with law and wilful blindness and infringement are also important to highlight. This book stems from a workshop that was organised in relation to a research project that explored the role that patents play in practice in the field of non-invasive prenatal testing.[3] Central to the findings of that project, and a previous project around genetic testing provision in the UK's National Health Service, were the ways in which patents were widely ignored, and the role that (non-)compliance plays in ensuring widespread access to patented technology.[4] The majority of

---

1 Limitations on the scope of patents which are challenged on more explicitly public interest bases in some jurisdictions may be imposed on corresponding patents in other jurisdictions through such means. Gene patents are such an example: Rochelle C Dreyfuss, Jane Nielsen and Dianne Nicol, 'Patenting Nature – A Comparative Perspective' (2018) 5 Journal of Law and the Biosciences 550; Naomi Hawkins, 'A Red Herring – Invalidity of Human Gene Sequence Patents' (2016) 38 European Intellectual Property Review 83; Naomi Hawkins, 'Human Gene Patents and Genetic Testing in Europe: A Reappraisal' (2010) 7 SCRIPTed 454.

2 Art 53(a) EPC and Rule 28(a)–(d) of the EPC 2000 Implementing Regulations.

3 The Impact of Patents on Translational Research – Non-Invasive Prenatal Diagnosis in Europe and the USA, UK Economic and Social Research Council, grant code ES/K009575/1.

4 Jane Kaye, Naomi Hawkins and Jenny Taylor, 'Patents and Translational Research in Genomics' (2007) 25 Nature Biotechnology 739; Naomi Hawkins, 'Human Gene Patents and Translational Research in Genetics' (DPhil thesis, University of Oxford 2010); Naomi Hawkins, 'The Impact of Human Gene Patents on Genetic Testing in the United Kingdom' (2011) 13 Genetics in Medicine 320; Naomi Hawkins, 'An Exception to Infringement for Genetic Testing – Addressing Patient Access and Divergence between Law and Practice' [2012] International Review of Intellectual

patent law scholarship does not explore these aspects, and is based instead on a more doctrinal analysis of law. But compliance, and wilful blindness towards patent law, also have important implications for our understanding of the way in which the public interest can be protected in this field. As highlighted in Chapter 2 of this book,[5] empirical evidence in patent law can play an important role in understanding the appropriate law and policy responses to emerging technologies in this area.

The fact remains that within patent law, the scope for explicit considerations of the public interest to play a central role is limited. Although these questions underpin existing law, they have been relegated by the interpretations of patent offices and courts to very narrow criteria.[6] Such narrow limits may be appropriate – they lend important certainty to the law, and avoid the judicial law-making that might otherwise result. However, it is a necessary result of such approach to interpretation that the role of the public interest has also been reduced. It is therefore right, as new technologies emerge, to question whether the current scope and interpretation of the law in this respect is sufficient, in light of those new technologies, to address potential concerns. Where the existing law and its application fails to sufficiently protect the public interest, then either it must be rebalanced, and tests adjusted and developed to take account of these interests, or other legal responses, such as in relation to defences to infringement or licensing rules, should be developed. Alternatively, policy responses, such as in relation to public funding of research or public procurement of drugs, may be a more targeted response.

In order to make such adjustments, scholars are increasingly arguing for greater attention to matters of the public interest by the institutions of patent law. This may include, for example, close attention to the requirements of patentability in this respect,[7] more attention to the role of the public in relation

---

Property and Competition Law 641; Naomi Hawkins, 'Patents and Non-Invasive Prenatal Testing: Is There Cause for Concern?' (2020) 47 Science and Public Policy 655.

[5] Andrew Sagar, Kathleen Liddell, Johnathon Liddicoat and Mateo Aboy, 'Patent Eligibility and Precision Medicine: Emerging Technologies, Legal Flux and Empirical Methods' in Naomi Hawkins (ed) *Patenting Biotechnological Innovation: Eligibility, Ethics and Public Interest* (Edward Elgar Publishing 2022).

[6] See for example the discussion of the exception to patentability of essentially biological processes in Karen Walsh, 'Invoking the Public Interest in the European Patent System' in Naomi Hawkins (ed) *Patenting Biotechnological Innovation: Eligibility, Ethics and Public Interest* (Edward Elgar Publishing 2022).

[7] Such as for example the requirement of non-obviousness: *Actavis Group v ICOS Corporation* [2019] UKSC 15; Justine Pila, 'Reflections on a Post-pandemic European Patent System' (2020) 42 European Intellectual Property Review 530.

to the patent opposition procedure[8] or a more expansive view of the morality provisions of patent law.[9] Some argue for private rights to be harnessed, such as through patent pledges[10] or patent pools.[11] Others argue for policy responses to address the arguable failings of the patent system to protect public interest. For example, initiatives such as the Covax initiative are designed to support the sharing of patented technology to ensure wider availability of Covid-19 vaccines and other technology.[12]

As is evident throughout the chapters of this book, the law works better to support the public interest in some areas than in others. For example, there is arguably flexibility in patentable subject matter, as evidenced by the development of the law around the patentability of personalised medicine,[13]

---

[8] Louise Hatherall, 'Procedural Issues in Public Interest Patent Challenges' (2022) 44 European Intellectual Property Review 198.

[9] Karen Walsh and Naomi Hawkins, 'Expanding the Role of Morality and Public Policy in European Patent Law' in Paul Torremans (ed), *Intellectual Property and Human Rights* (Wolters Kluwer 2020); Geertrui van Overwalle, 'Human Rights', Limitations in Patent Law' in FW Grosheide (ed), *Intellectual Property and Human Rights: A Paradox* (Edward Elgar Publishing 2010).

[10] Jorge L Contreras and others, 'Pledging Intellectual Property for COVID-19' (2020) 38 Nature Biotechnology 1146; Karen Walsh and others, 'Intellectual Property Rights and Access in Crisis' (2021) 52 IIC International Review of Intellectual Property and Competition Law 379.

[11] G van Overwalle and others, 'Models for Facilitating Access to Patents on Genetic Inventions' (2006) 7 Nature Reviews Genetics 143; Carl Shapiro, 'Navigating the Patent Thicket: Cross Licenses, Patent Pools and Standard Setting' in Adam B Jaffe, Josh Lerner and Scott Stern (eds), *Innovation Policy and the Economy* (National Bureau of Economic Research, Innovation Policy and the Economy Group, MIT Press 2001); Geertrui van Overwalle, *Gene Patents and Collaborative Licensing Models: Patent Pools, Clearinghouses, Open Source Models, and Liability Regimes* (Cambridge University Press 2009).

[12] World Health Organization, 'COVAX' www.who.int/initiatives/act-accelerator/covax accessed 7 November 2021; Lisa M Herzog and others, 'Covax Must Go beyond Proportional Allocation of Covid Vaccines to Ensure Fair and Equitable Access' (2021) 372 BMJ.

[13] Andrew Sagar, Kathleen Liddell, Johnathon Liddicoat, Mateo Aboy, 'Patent Eligibility and Precision Medicine: Emerging Technologies, Legal Flux and Empirical Methods' in Naomi Hawkins (ed) *Patenting Biotechnological Innovation: Eligibility, Ethics and Public Interest* (Edward Elgar 2022).

non-invasive prenatal testing[14] and biofabrication[15] to allow for at least some consideration of the balance of private rights and the public interest. However, international divergence in approaches can give rise to some degree of fragmentation, raising concerns for the impact on business. An excessively narrow approach to the interpretation of exceptions, exclusions and defences can result in a broadening of the scope of patentability, with a consequent narrowing of the scope for the invocation of the public interest.[16] The morality provisions have come under a great deal of scrutiny and have at times been criticised for being interpreted too narrowly, inconsistently or incoherently, and not serving to adequately safeguard the public interest as a result.[17] Important scope therefore remains for a more coherent and effective protection of the public interest when having regard to the initial grant and validity of a patent.

Beyond the grant and validity of patents, the exploitation of patents is central to the way in which the public interest is protected. Voluntary licensing is the main means through which patents are exploited, and therefore has an important impact. The private rights, freedom of contract and lack of public scrutiny of the majority of licensing agreements mean that the public interest is entrusted to the freedom of the market, except for when licensing relationships break down, or where licences cannot be agreed through negotiation. Compulsory licences have played at least a theoretical role in patents for many years, although the rareness of their exercise suggests that any effectiveness arises largely as a result of their threatened application, which spurs patent holders to come to a voluntary agreement.[18] Increasing attention is being paid

---

[14] Dianne Nicol and Jane Nielsen, 'Non-invasive Prenatal Testing and the Resilience of the Patent System' in Naomi Hawkins (ed) *Patenting Biotechnological Innovation: Eligibility, Ethics and Public Interest* (Edward Elgar Publishing 2022); Naomi Hawkins and others, 'The Continuing Saga of Patents and Non-Invasive Prenatal Testing' (2019) 39 Prenatal Diagnosis 44.

[15] Pheh Hoon Lim and Phoebe Li, 'Patentability of Biofabricated Human Organs: "Products of Nature" or "Products Derived from Nature" Revisited' in Naomi Hawkins (ed) *Patenting Biotechnological Innovation: Eligibility, Ethics and Public Interest* (Edward Elgar Publishing 2022); Phoebe H Li, '3D Bioprinting Technologies: Patents, Innovation and Access' (2014) 6(2) Law Innovation and Technology 282.

[16] See for example the discussion in Walsh (n 6).

[17] See for example Ella O'Sullivan, 'A Problem of Interpretation: Is Article 53(a) EPC Necessarily Restrictive of Emerging Technologies?' in Naomi Hawkins (ed) *Patenting Biotechnological Innovation: Eligibility, Ethics and Public Interest* (Edward Elgar Publishing 2022); Walsh and Hawkins (n 9); Justine Pila, 'Adapting the Ordre Public and Morality Exclusion of European Patent Law to Accommodate Emerging Technologies' (2020) 38 Nature Biotechnology 555.

[18] For a more detailed discussion, see: Walsh and others (n 10); WR Cornish, M Llewelyn and M Adcock, 'Intellectual Property Rights (IPRs) and Genetics: A Study into the Impact and Management of Intellectual Property Rights within the Healthcare

to licensing, such as through the Fair, Reasonable and Non-Discriminatory (FRAND) licence negotiations for standard essential patents, where questions of the public interest are scrutinised.[19] Moreover, the possibility of leveraging private rights in the public interest, to ensure ethical licensing, are also worth exploring.[20] Increased attention to the ways in which the public interest can be foregrounded in the exploitation of patents, particularly where the patents arise from publicly funded research or out of publicly supported infrastructures such as biobanks and collaborative platforms in genomics, is necessary.[21] Such attention, to allow purely private interests to be ameliorated to ensure that the patent bargain is being met at this end of the patent lifespan, is essential

Protection of the public interest is vital to patent law, especially in relation to biotechnology. The limits of patent law in addressing the intractable problems in this area have been clear in relation to questions of access to healthcare for some time, unambiguously highlighted by the HIV/AIDS crisis[22] and more recently in the patent-related concerns around Covid-19.[23] Similarly, climate

---

Sector' (Public Health Genetics Unit, Cambridge Genetics Knowledge Park 2003); Lasantha Aritarathna and Kanchana Kariyawasam, 'Pharmaceutical Patents and Access to Generic Medicines in Developing Countries' (2020) 42(2) EIPR 108, 111–14; Esther van Zimmeren and Geertrui Van Overwalle, 'A Paper Tiger? Compulsory License Regimes for Public Health in Europe' (2011) 42 IIC 4.

19 See for example Alison Slade, 'Repositioning the Access to Healthcare Debate: FRAND and the Internet of Medical Things' in Naomi Hawkins (ed) *Patenting Biotechnological Innovation: Eligibility, Ethics and Public Interest* (Edward Elgar Publishing 2022); Jorge L Contreras, 'From Private Ordering to Public Law: The Legal Frameworks Governing Standards-Essential Patents' (2017) 30 Harvard Journal of Law and Technology 211, 227.

20 Aisling McMahon, 'Accounting for Ethical Considerations in the Licensing of Patented Biotechnologies and Health-Related Technologies: A Justification' in Naomi Hawkins (ed) *Patenting Biotechnological Innovation: Eligibility, Ethics and Public Interest* (Edward Elgar Publishing 2022).

21 Hermann Garden, Naomi Hawkins and David Winickoff, (2021) 'Building and Sustaining Collaborative Platforms in Genomics and Biobanks for Health Innovation', OECD Science, Technology and Industry Policy Papers, No. 102, OECD Publishing, Paris, https://doi.org/10.1787/11d960b7-en.

22 Ellen T'Hoen and others, 'Driving a Decade of Change: HIV/AIDS, Patents and Access to Medicines for All' (2011) 14 Journal of the International AIDS Society 15; Carolyn Deere Birkbeck, *The Implementation Game: The TRIPS Agreement and the Global Politics of Intellectual Property Reform in Developing Countries* (Oxford University Press 2008).

23 Sivaramjani Thambisetty, Aisling McMahon, Luke Mcdonagh, Hyo Yoon Kang and Graham Dutfield, 'The TRIPS Intellectual Property Waiver Proposal: Creating the Right Incentives in Patent Law and Politics to End the COVID-19 Pandemic' LSE Legal Studies Working Papers (06/2021). London School of Economics and Political Science, London, UK; 'A Patent Waiver on COVID Vaccines Is Right and Fair' (2021) 593 Nature 478; Aisling McMahon, 'Global Equitable Access to Vaccines, Medicines

change and the role that patents might play in ensuring access to climate technologies is an area which seems likely to grow in importance in the near future.[24] Where there are such important interests at stake, the rights of private individuals must be appropriately balanced against the rights of the wider public to benefit from innovations which have the capacity to make an important difference to public welfare, whether that is in relation to diseases such as HIV or malaria, which have had a devastating impact in the global south; to pandemic diseases such as Covid-19; or to climate change. It is right, and indeed essential, that the role that patent law plays in incentivising innovation in these areas, as well as in restricting access to those innovations, is closely examined.

What should be evident from this book, as well as from wider scholarship in the field, is that the questions of public interest in this field must be addressed in multiple ways – there is no single magic bullet. It is therefore important that the limits and scope of patentable subject matter are scrutinised and tested, such as for example in relation to patents on human genes,[25] or through the examination of the limits of method patents for diagnosis.[26] Moreover, the scope and extent of limitations on patents as set out in the exceptions to patentability, such as for *ordre public* and morality, must be explored, and a consist-

---

and Diagnostics for COVID-19: The Role of Patents as Private Governance' (2021) 47 Journal of Medical Ethics 142; Aisling Mcmahon, 'Patents, Access to Health and COVID-19 – The Role of Compulsory and Government-Use Licensing in Ireland' (2020) 71 Northern Ireland Legal Quarterly 331.

[24] See discussion in Abbe Brown, *Intellectual Property, Climate Change and Technology: Managing National Legal Intersections, Relationships and Conflicts* (Edward Elgar Publishing 2019).

[25] *Association for Molecular Pathology v Myriad Genetics Inc* 569 US 576 (2013); *D'Arcy v Myriad Genetics Inc* [2015] HCA 35; Dianne Nicol and others, 'International Divergence in Gene Patenting' (2019) 20 Annual Review of Genomics and Human Genetics 519; Dreyfuss, Nielsen and Nicol (n 1).

[26] *Mayo Collaborative Services v Prometheus Laboratories Inc* 566 US 66 (2012); *Association for Molecular Pathology v Myriad Genetics Inc* 569 US 576 (2013); *Alice Corp v CLS Bank International* 573 US 208 (2014); *Vanda Pharmaceuticals Inc v West-Ward Pharmaceuticals International Ltd* in F.3d Vol. 887 1117 (United States Court of Appeals for the Federal Circuit, 2018); Andrew Sagar, Kathleen Liddell, Johnathon Liddicoat and Mateo Aboy, 'Patent Eligibility and Precision Medicine: Emerging Technologies, Legal Flux and Empirical Methods' in Naomi Hawkins (ed) *Patenting Biotechnological Innovation: Eligibility, Ethics and Public Interest* (Edward Elgar Publishing 2022); Dianne Nicol and Jane Nielsen, 'Non-invasive Prenatal Testing and the Resilience of the Patent System' in Naomi Hawkins (ed) *Patenting Biotechnological Innovation: Eligibility, Ethics and Public Interest* (Edward Elgar Publishing 2022); M Aboy, C Crispo, K Liddell, T Minssen and J Liddicoat, 'Mayo's Impact on Patent Applications Related to Biotechnology, Diagnostics and Personalized Medicine' (2019) 37 Nature Biotechnology 513–19.

ent, coherent basis formed. Finally, scope for carving out responses to protect the public interest in the exploitation of patents, whether through competition law means such as FRAND[27] or through the exercise of private rights by actors such as research funders who are in a position to assert their bargaining power to defend the public interest, must also be explored.

The public interest is too important to ignore, and it is therefore vital that scholars and policy makers turn their attention to its role and draw on the existing mechanisms within the law for its safeguard. At the same time, it is also important that the limitations of patent law be recognised in this regard. Solutions which are grounded solely in the realm of patent law are insufficient to address the intractable problems of access to innovation. Instead it is necessary to look beyond the pure patent law sphere, to areas such as, but not limited to, contract and competition law, to support the important issues at stake. Recourse to policy and public procurement also can provide important avenues for solutions which can work hand in hand with the areas outlined in this book. A more cohesive approach to these questions, which both draws on patent law and looks beyond its narrow provisions to mechanisms of other areas of law and policy, can only work to safeguard the public interest more securely. This will then in turn serve to strengthen patent law's justifications overall, balancing the private right of the inventor with the public interest. An appropriately balanced patent law will enhance public trust in this area of law, which will in turn support patent law to play an appropriate and important role in the future of innovation.

---

[27] See discussion in Pila (n 7).

Index